Decomposition Methods for Differential Equations

Theory and Applications

CHAPMAN & HALL/CRC
Numerical Analysis and Scientific Computing

Aims and scope:

Scientific computing and numerical analysis provide invaluable tools for the sciences and engineering. This series aims to capture new developments and summarize state-of-the-art methods over the whole spectrum of these fields. It will include a broad range of textbooks, monographs and handbooks. Volumes in theory, including discretisation techniques, numerical algorithms, multiscale techniques, parallel and distributed algorithms, as well as applications of these methods in multi-disciplinary fields, are welcome. The inclusion of concrete real-world examples is highly encouraged. This series is meant to appeal to students and researchers in mathematics, engineering and computational science.

Proposals for the series should be submitted to one of the series editors above or directly to:
CRC Press, Taylor & Francis Group
4th, Floor, Albert House
1-4 Singer Street
London EC2A 4BQ
UK

Published Titles

A Concise Introduction to image Processing using C++
Meiqing Wang and Choi-Hong Lai

**Decomposition Methods for Differential Equations:
   Theory and Applications**
Juergen Geiser

**Grid Resource Management: Toward Virtual and Services Compliant
Grid Computing**
Frédéric Magoulès, Thi-Mai-Huong Nguyen, and Lei Yu

Introduction to Grid Computing
Frédéric Magoulès, Jie Pan, Kiat-An Tan, and Abhinit Kumar

Numerical Linear Approximation in C
Nabih N. Abdelmalek and William A. Malek

Parallel Algorithms
Henri Casanova, Arnaud Legrand, and Yves Robert

Parallel Iterative Algorithms: From Sequential to Grid Computing
Jacques M. Bahi, Sylvain Contassot-Vivier, and Raphael Couturier

Decomposition Methods for Differential Equations

Theory and Applications

Juergen Geiser

CRC Press
Taylor & Francis Group
Boca Raton   London   New York

CRC Press is an imprint of the
Taylor & Francis Group, an **informa** business

A CHAPMAN & HALL BOOK

CRC Press
Taylor & Francis Group
6000 Broken Sound Parkway NW, Suite 300
Boca Raton, FL 33487-2742

First issued in paperback 2017

© 2009 by Taylor & Francis Group, LLC
CRC Press is an imprint of Taylor & Francis Group, an Informa business

No claim to original U.S. Government works

ISBN 13: 978-1-138-11414-2 (pbk)
ISBN 13: 978-1-4398-1096-5 (hbk)

Visit the Taylor & Francis Web site at
http://www.taylorandfrancis.com

and the CRC Press Web site at
http://www.crcpress.com

Preface

The monograph was written at the Department of Mathematics, Humboldt-Universität zu Berlin, Unter den Linden 6, D-10099 Berlin, Germany.

In the monograph we discuss the decomposition methods with respect to the discretisation method for evolution equations. The considered evolution equations are systems of parabolic and hyperbolic equations. We deal with convection-diffusion-reaction equations, heat-equations and wave-equations. The applications are transport-reaction models, heat transfer models, micromagnetics and elastic-wave propagation models.

We propose efficient time and spatial decomposition methods, with respect to the underlying discretization and solver methods. The main advantage is efficiency in the computational- and memory-resources based on the simplification in the decoupled equations. Further we propose the combination of time and spatial decomposition and discretization methods, so that the splitting and discretization error can be neglected with respect to the higher order methods. One benefit of the iterative splitting method is to have all operators in their iteration steps, and such behavior preserves the physics in each equations.

We discuss in the theoretical part the stability and consistency of the decomposition methods and in the practical part the application of the methods with their benefits in decomposing test- and real-life-problems.

This work has been accompanied and supported by many coworkers and colleagues and I would like to thank all of them.

First, I gratefully want to thank my colleagues and coworkers at the Humboldt-University of Berlin for their fruitful discussions and ideas. In particular, I wish to thank Prof.Q. Sheng, Prof.I. Farago, Dr.R. Müller and the student workers L. Noack, V. Schlosshauer and R. Huth for their support and ideas.

My special thanks to my wife Andrea and my daughter Lilli who have always supported and encouraged me.

Berlin, December 2008 Jürgen Geiser

Contents

List of Figures

List of Tables

Introduction

In this monograph, we describe the analysis of numerical methods for evolution equations that are based on temporal and spatial decomposition methods. The decomposition methods are discussed with respect to their effectivity, combination possibility with discretization methods, multi scaling possibilities, and stability to initial and boundary values problems. The treatment of spatial-dependent, stiff, and nonlinear operators are discussed, because such effects are incorporated in our proposed multi-physics model problems.

These multi-physics problems are presented respecting the different physical behaviors for each simpler discretized and decoupled equation. The decomposition is described on how to decouple the problems efficiently without losing the physical correctness. The discretization is described on how to approximate the problems accurately without loosing the physics. The aim is to achieve simpler differential equations and to compute with higher order discretization and solver methods.

Here the mathematical contribution is decomposing to simpler models, which can be handled with effective solver methods. The following contributions are discussed in the monograph:

- Generalize the numerical analysis with respect to the consistency and stability to nonlinear, stiff, and spatial decomposed splitting problems

- Adapt the discretization to the decomposition methods,

- Efficient decomposition and discretization methods, with respect to computational time and memory

- Embed higher-order time-discretization methods to the decoupled equations

- Apply the results in computational sciences (e.g., flow problems, elastic wave propagation, heat transfer, magnetic problems)

The work presented in the next chapters is related to real-life problems, the underlying decomposition and discretization, the stability and consistency analysis of the proposed decomposition methods, and the numerical results.

First we present the modeling of selected multi-physics problems. We describe the underlying characteristics of the various equation parts and their spatiotemporal behavior. The knowledge allows us to design special decoupling methods and to respect the underlying conservation of physics, see also applications in [118].

In the next parts we describe the discretization and decomposition to achieve decoupled equations for the time- and space-splitting methods. We discuss the discretization for the evolution equations to obtain the abstract operator equations.

The advantages of time- and space-decomposition methods are presented as the possibility to adapt the best discretization and solver method for the decomposed problems. The order of large problems is reduced into simpler and efficiently solvable partial problems either in space or in time.

For the discretization methods we discuss the spatial discretization methods of Finite Volume and Discontinuous Galerkin methods, which can be applied for conservation laws. Further, we take into account that the decomposition methods have their benefits in decoupling the appropriate operators, with respect to their physical and mathematical behavior, see [82] and [95]. For these we discuss the scales of each operator and collocate the operators with respect to these scales.

We propose, based on this assumption, for the temporal discretizations higher-order Runge-Kutta methods and for the spatial discretization higher-order finite-volume and finite-element methods and also mixed discretization methods with embedded local analytical solutions.

For our main contribution to the decomposition methods, the benefits to the classical time-splitting methods such as operator-splitting methods are explained. An improvement of such methods to higher methods is given in the form of an intensive analysis of the iterative operator-splitting method.

In the next step the extension to hyperbolic equations is discussed as a further application of the iterative operator-splitting methods. Due to a reduction to first-order systems of ordinary differential equations, the stability and consistency analysis can be embedded to the theory of the splitting methods for parabolic equations.

Therefore, a closed error analysis of the iterative operator-splitting methods can be presented for the evolution equation with respect to their stiffness, spatiality, and nonlinearity. All in all, the new iterative operator-splitting methods have the advantage as decomposition methods to obtain higher-order methods and to embed nonlinearities also as spatial-splitting methods, see [88], [109], and [155].

The contents of the monograph are presented in eight chapters and cover a large range of discretizations, decompositions, and applications arising from multi physical and multi dimensional problems.

Chapter 1 is the introduction to the monograph.

In Chapter 2 we introduce the decomposition analysis. The decomposition idea is proposed and a brief introduction is given to the underlying discretization methods. The aim is to obtain systems of ordinary differential equations, presented as operator equations, which can be used in the decomposition methods.

In Chapter 3 we discuss the time and space discretization methods for the parabolic and hyperbolic equations. The aim is to obtain a semi-discretization that can be applied by the decomposition methods.

In Chapter 4 we describe the temporal decomposition methods to the parabolic equations. The operator-splitting methods are discussed with respect to the computational efficiencies and the multi scaling properties. We present recent results to this topic and we take into account the contributions in stability and consistency analysis for systems of partial differential equations. Moreover, the theory of the iterative operator-splitting methods is described and the connection with other splitting methods, as the classical sequential splitting methods, is presented. Extended results in the direction of quasilinear systems of parabolic equations are discussed, and the stability results for weighted methods are presented.

In Chapter 5 we describe the temporal decomposition methods to the hyperbolic equations. The higher-order derivatives in time are discussed with the application to operator-splitting methods. First the classical alternating direction implicit (ADI) and locally one-dimensional (LOD) methods are presented and then the iterative operator-splitting methods extended to the second-order time derivatives. Recent results to this topic are taken to account, and we present the contributions in stability and consistency analysis for systems of second-order differential equations. Moreover, the theory of the iterative operator-splitting methods is described, and the connection with other splitting methods, as the ADI, LOD methods, is presented. Further, the parallelization of the splitting methods is discussed.

In Chapter 6 we discuss the combination of time and spatial decomposition methods. Such mixed methods can be taken into account for weak coupled systems of differential equations.

In Chapter 7 we present the application of our methods to numerical problems. We present test examples and real-life problems in physical and engineering applications. The test examples verified our theoretical results of the splitting methods and showed the convergence. The real-life problems are proposed in several applications. So a first application is done for transport-reaction processes in a waste disposal. The simulations show results in the contamination of the underlying rock. Further, the heat process of a growth apparatus for crystal-growth processes is argued to obtain results about the temperature in the gas chamber, see [96]. We apply our methods to hyperbolic problems as the elastic wave propagation model. In such a model we simulate simple earthquakes and obtain results about the strength of the ground forces, see [55] and [56].

Computational results are computed with different software tools such as MATLAB, R^3T (software product developed at the University of Heidelberg, Germany), WIAS-HiTNIHS (software product developed by the Weierstrass Institute, Berlin, Germany), and $OPERA - SPLITT$ (software tool based on MATLAB, developed at the Humboldt Universität zu Berlin, Germany).

In Chapter 8 we summarize our results of the monograph and proposed some future works.

In the Appendix we present the notation of our abbreviations, the literature,

4

and the index.

Chapter 1

Modeling: Multi-Physics Problems

1.1 Introduction

In this chapter we introduce our monograph and the related mathematical theory and applications.

In various applications in the material physics, geoscience, and chemical engineering, the simulations of multi-physics problems are very important.

A multi-physics problem is defined as a problem with different physical processes, dependent in time and space (e.g., flow-process, reaction-process, growth-process, etc.).

Because of the large differential equation systems, an enormous increase in performance for the simulation programs is necessary and high complex algorithms are needed for fast computations.

The mathematical models of the multi-physics problems consist usually of coupled partial differential equations. They reflect the underlying physical behavior of the processes. The most relevant problems are no longer analytically solvable and so numerical solutions are necessary.

Based on the physical behavior, the numerical methods are developed in such a manner that the physical laws are conserved.

We will concentrate on the multi-physics problems based on linear and non-linear coupled parabolic and hyperbolic equation systems. Thus, the coupled equation systems have at least parabolic characteristics.

The questions in this monograph are how to decouple the equations to conserve the physical behavior and how to accelerate the solver process for simpler equations and how to achieve higher-order methods for more accurate computations.

In the next sections we discuss the multi-physics problems.

1.2 Models for Multi-Physics Problems

We will concentrate on a family of multi-physics problems related to flow and reaction problems, where we separate the flow problems into convection,

5

diffusion and wave-propagation problems, see [23], [24], [41], [60], [61] and [66].

The underlying model is funded in applications for transport reaction, heat, and flow processes in geophysics, material sciences, quantum physics, and more.
The problem can be discussed for weak- and strong-coupled equations, thus for each physical process we can concentrate on simpler model equations that are sufficient for analyzing the multi-physics problems.

1.3 Examples for Multi-Physics Problems

The next problems are selected to apply our decomposition methods and to discuss the influence of the development of the splitting ideas, see [193].

1.3.1 Waste Disposal

We simulate a waste disposal for radioactive waste in a salt dome. The salt dome is surrounded by a spacious, heterogenous overlaying rock (a schematical overview is presented in Figure 1.1).
For a potential waste case, groundwater spools in the possible pathway inside the waste disposal and contacts with radioactive waste.
Because of the high pressures and motions in the salt dome, the overlaying rock presses out the contaminant water for the salt dome. The contaminant water is then used as a time-dependent source for the radionuclides in the groundwater flow.
The radionuclides are transported via the groundwater and we neglect the pressed out water mass compared with the water of the groundwater flow. For the reactive processes, we use the radioactive decay that denotes the transfer from a radionuclide in the corresponding child nuclide and the adsorption that denotes the exchange in the several abidance areas for the mobile or immobile phase, cf. [81].
The modeling for these cases is done in [24], [47], [135], and [73]. Based on the physical model, we could derive the mathematical models.
We describe for simplification a model for equilibrium sorption. This model is studied in various directions and we explain the model in [81]. The phases for the contaminants are presented in the mobile, immobile, sorption, and immobile sorption phase, cf. [81].

The model equations are given as

$$\phi \, \partial_t R_i c_i + \nabla \cdot (\mathbf{v} c_i - D \nabla c_i) = -\phi R_i \, \lambda_i c_i + \sum_{k=k(i)} \phi \, R_k \, \lambda_k c_k + \tilde{Q}_i, \quad (1.1)$$

$$c_{e(i)} = \sum_i c_i , \quad R_i = 1 + \frac{(1-\phi)}{\phi} \rho \, K(c_{e(i)}), \quad \text{with} \quad i = 1, \ldots, M,$$

where c_i is the i-th concentration, R_i is the i-th retardation factor, λ_i the i-th decay constant, and $\mathbf{v}$ the velocity field that is computed by another program package (e.g., $\mathbf{d^3f}$-program package, see [72]) or that is given a priori. Q_i is the i-th source-term, ϕ is the porosity, $c_{e(i)}$ is the sum of all isotopic concentrations of the element e, K is a function of the isotherms, see [81]. M is the number of the radioactive concentrations.

Figure 1.1 presents the physical circumstances of a waste disposal as given in the task.

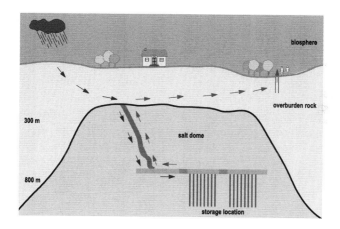

FIGURE 1.1: Schematical overview of the waste disposal, cf. [32]

REMARK 1.1 For this model the splitting methods are applied with respect to decouple the fast reaction processes and the slow transport processes. We can apply fast ordinary differential equation (ODE) solvers for the

reaction processes and implicit partial differential equation (PDE) solvers for
the transport processes, see [124]. ⬜

1.3.2 Crystal Growth

The motivation for modeling an accurate technical apparatus or physical
process is coming from the demand to have a tool for developing an optimal
and efficient apparatus for foreseeing the physical effects and protecting the
environment.

The basic idea for the model is the transfer between reality and the possible
abstraction for an implementable model. Often some interested effects are
sufficient to develop a simpler model from the reality.

Silicon carbide (SiC) is a wide-bandgap semiconductor used in high-power
and high-frequency industrial applications: SiC serves as substrate material
for electronic and optoelectronic devices such as MOSFETs, thyristors, blue
lasers, and sensors (see [152] for a recent account of advances in SiC devices).
Its chemical and thermal stability make SiC an attractive material to be used
in high temperature applications as well as in intensive radiation environ-
ments. For an economically viable industrial use of SiC, growth techniques
for large diameter, low defect SiC boules must be available. Recent years
have seen steady improvement (see [122]) of size and quality of SiC single
crystals grown by sublimation via *physical vapor transport* (PVT, also known
as modified Lely method, see, e.g., [140]). However, many problems remain,
warranting further research.

Typically, modern PVT growth systems consist of an induction-heated
graphite crucible containing polycrystalline SiC source powder and a single
crystalline SiC seed (see Figure 6.19). The source powder is placed in the hot
zone of the growth apparatus, whereas the seed crystal is cooled by means of a
blind hole, establishing a temperature difference between source and seed. As
the SiC source is kept at a higher temperature than the cooled SiC seed, sublimation is encouraged at the source and crystallization is encouraged at the
seed, causing the partial pressures of Si, Si_2C, and SiC_2 to be higher in the
neighborhood of the source and lower in the neighborhood of the seed. As the
system tries to equalize the partial pressures, source material is transported
to the seed which grows into the reaction chamber.

Because of the complex processes, a careful study is important to correctly
design the numerical simulations, [165]. Based on this background the combination of discretization and solver methods is an important task. We propose
the decomposition methods of breaking down complicated multi-physics in
simpler physics. The time-decomposition methods and their extended versions with more stabilized behavior are based on operator-splitting methods,
see [70]. With these methods a useful decoupling of the time scales is possible, and the solvers can be applied on the different time scales. Further, the
space-decomposition methods are based on the Schwarz waveform-relaxation
methods and their accurate error estimates, see [54]. The methods decouple

into domains with the same equation parameters, therefore effective spatial discretization and solver methods are applicable.

The model is given as follows:

a) We assume that the temperature evolution inside the gas region Ω_g can be approximated by considering the gas as pure argon. The reduced heat equation is given as

$$\rho_g \partial_t U_g - \nabla \cdot (\kappa_g \nabla T) = 0, \qquad (1.2)$$

$$U_g = z_{Ar} R_{Ar} T, \qquad (1.3)$$

where T is the temperature, t is the time, and U_g is the internal energy of the argon gas. The parameters are given as ρ_g being the density of the argon gas, κ_g being the thermal conductivity, z_{Ar} being the configuration number, and R_{Ar} being the gas constant for argon.

b) The temperature evolution inside the region of solid materials Ω_s (e.g., inside the silicon carbide crystal, silicon carbide powder, graphite, and graphite insulation), is described by the heat equation:

$$\rho_s \partial_t U_s - \nabla \cdot (\kappa_s \nabla T) = f, \qquad (1.4)$$

$$U_s = \int_0^T c_s(S) \, dS, \qquad (1.5)$$

where ρ_s is the density of the solid material, U_s is the internal energy, κ_s is the thermal conductivity, and c_s is the specific heat. f represents the heat source in the material Ω_s.

The equations hold in the domains of the respective materials and are coupled by interface conditions, for example, those requiring the continuity for the temperature and for the normal components of the heat flux on the interfaces between opaque solid materials. On the boundary of the gas domain (i.e., on the interface between the solid material and the gas domain), we consider the interface condition

$$\kappa_g \, \nabla T \cdot \mathbf{n}_g + R - J = \kappa_s \, \nabla T \cdot \mathbf{n}_g, \qquad (1.6)$$

where $\mathbf{n}_g$ is the normal vector of the gas domain, R is the radiosity, and J is the irradiosity. The irradiosity is determined by integrating R along the whole boundary of the gas domain, see [136]. Moreover, we have

$$R = E + J_{\text{ref}}, \qquad (1.7)$$

$$E = \sigma \, \epsilon \, T^4 \quad \text{(Stefan-Boltzmann equation)}, \qquad (1.8)$$

$$J_{\text{ref}} = (1 - \epsilon) \, J, \qquad (1.9)$$

where E is the radiation, J_{ref} is the reflexed radiation, ϵ is the emissivity, and σ is the Boltzmann radiation constant.

REMARK 1.2 The splitting methods are applied with respect to the anisotropy in the different dimensions. Therefore, a dimensional splitting can accelerate the solver process and save memory resources, see [86] and [155]. ☐

In the next section, we focus on elastic wave propagation.

1.3.3   Elastic Wave Propagation

The motivation to study the elastic wave propagation came from the earthquake simulation. The realistic earthquake sources and complex three-dimensional (3D) earth structure are of immense interest in understanding the earthquake formation. A foundation for model ground motion in urban sedimentary basins are studied and 3D software-packages are developed, see [55] and [56]. Numerical simulations of wave propagation can be done in two and three dimensions for models with sufficient realism (e.g., 3D geology, propagating sources, frequencies approaching 1 Hz) to be of engineering interest. Before numerical simulations can be applied in the context of engineering studies or seismic hazard analysis, the numerical methods and the models associated with them must be thoroughly validated, in this process also the splitting methods are involved. Further the interest on simulating accurate ground motion from propagating earthquakes in 3D earth models is important. We propose the first higher-order splitting method in this context and have done the first 3D simulations of simpler earthquake models to test the accuracy of split equations and the conservation of the physical effects. The model problem is given with the physical parameters for the underlying domains in which the earthquake occurs.

We present the model problem as an elastic wave equation for constant coefficients in the following notations:

$$\rho \partial_{tt} \mathbf{U} = \mu \nabla^2 \mathbf{U} + (\lambda + \mu) \nabla (\nabla \cdot \mathbf{U}) + \mathbf{f}, \qquad (1.10)$$

where $\mathbf{U}$ is equal to $(u, v)^T$ or $(u, v, w)^T$ in two and three dimensions, and $\mathbf{f}$ is a forcing function. In seismology it is common to use spatial singular forcing terms, which can look like

$$\mathbf{f} = \mathbf{F} \delta(\mathbf{x}) g(t), \qquad (1.11)$$

where $\mathbf{F}$ is a constant direction vector. A numeric method for Equation (1.10) needs to approximate the Dirac function $\delta(\mathbf{x})$ correctly in order to achieve full convergence.

The contribution to this model is the initial process and spatial dependent processes, which are started as a singular function (Dirac function) and are computed in the later time-sequences as smooth functions. Therefore, large time-steps are possible in the computation to simulate the necessary time-periods.

REMARK 1.3 Here the splitting methods are applied with respect to the later time-sequences to guarantee large time-steps. The different spatial dependent processes can be decoupled into Laplacian operators and non-Laplacian operators. Such effective decoupling allows us to accelerate the solver-process of the simpler equations and save computational resources, see [56] and [103]. □

1.3.4 Magnetic Trilayers

The motivation for the study is coming from modeling ferromagnetic materials, used in data storage devices and laptop displays. The models are based on the Landau-Lifschitz-Gilbert equation, see [144], and are extended by the interaction of different magnetic layers. The application of such model problems includes switch processes in magnetic transistors (FET) or data devices. Mathematically the equations cannot be solved analytically and numerical methods are important. Therefore, numerical methods for stable discretizations are important because of the known blow-up effects, see [19], [174]. We apply the scalar theory and present the extention to more complicate systems of magnetic models. In the numerical examples we present first results of single layers and influence with external magnetic fields.

In the mathematical model we deal with magnetic multi layers described by the coupled Landau-Lifschitz-Gilbert equation.

The Landau-Lifschitz free energy E for the 2 layers is given as

$$E(\mathbf{m}_1) = \int_\omega \left(1/2|\nabla \mathbf{m}_1|^2 + \phi(\mathbf{m}_1)\right) \tag{1.12}$$
$$+\langle \nabla u, \mathbf{m}_1 \rangle_{\mathbb{R}^2} - \langle h_{ext,1}, \mathbf{m}_1 \rangle + A(\mathbf{m}_1, \mathbf{m}_2)) \, dx$$

$$E(\mathbf{m}_2) = \int_\omega \left(1/2|\nabla \mathbf{m}_2|^2 + \phi(\mathbf{m}_2)\right) \tag{1.13}$$
$$+\langle \nabla u, \mathbf{m}_2 \rangle_{\mathbb{R}^2} - \langle h_{ext,2}, \mathbf{m}_2 \rangle + A(\mathbf{m}_1, \mathbf{m}_2)) \, dx$$

$$A(\mathbf{m}_1, \mathbf{m}_2) = \sum_{k=1}^{2} A_k \langle \mathbf{m}_1, \mathbf{m}_2 \rangle \tag{1.14}$$

where $\mathbf{m}_i$ is the magnetization vector in the layer i ($i \in \{1, 2\}$), $\gamma_i' > 0$, the saturation magnetization is given as $M_{s,i}$ for $i \in \{1, 2\}$. $h_{ext,i}$ is the external field of layer $i \in \{1, 2\}$.

For the magnetostatic case, the magnetic potential u_i and the magnetization $\mathbf{m}_i$ are related through

$$\Delta u_i = div(\xi_\omega \mathbf{m}_i) \text{ in } \mathbb{R}^2, \tag{1.15}$$

where $\omega \subset \mathbb{R}^2$ is the domain covered by the ferromagnet, and $\xi_\omega = 1$ on ω and 0 else.

The magnetic potential is also called a dipole field.

REMARK 1.4 Here the splitting methods are applied with respect to the different layers. The physical decomposition into single-layer models is performed. Such effective decoupling allows us to accelerate the solver-process of the physical simpler equations, see [19]. ⬜

In the next chapter we discuss the decomposition analysis and discretization methods to obtain the operator equations.

Chapter 2

Abstract Decomposition and Discretization Methods

In this chapter we briefly introduce the underlying ideas of the decomposition analysis and discretization methods to obtain the abstract operator equations, which are used and studied for our time decomposition methods.

2.1 Decomposition

In this section we discuss the decomposition of the evolution equations to obtain separate underlying equations. The simpler equation parts are discretized to obtain an abstract operator equation formulation. Based on this formulation we apply the decomposition methods, see Chapters 3 and 4 for detailed explanations.

In the decomposition of the evolution equations, we deal with two main problems:

- Decomposition of the evolution equation

- Decomposition methods to solve the decoupled evolution equation

The decomposition can be performed before the discretization, in which case the physical behaviors of the operators are used to decide the type of the splitting operators. Further, the decomposition can be done after the space and time discretization of the partial differential equation, such that the spectral analysis or stability conditions can be used to decide whether operators are initiated for the splitting process.

We focus on the decomposition methods, which are applied after the space discretization, and we obtain a system of ordinary differential equations. On this level, the consistency and stability analysis can be treated on an abstract level as a Cauchy problem in a more abstract level.

In the next subsection, we consider the decomposition of the evolution equations and present two methods: the direct method called the *physical decomposition method* and the more indirect method called the *mathematical decomposition method*.

2.1.1 Decomposition of the Evolution Equations

The first step of decomposing an equation into simpler problems is to question the temporal, spatial, and physical scales of the equations.

In our work we contribute the temporal scales and propose two possible mechanisms for decomposing an evolution equation:

- Physical decomposition

- Mathematical decomposition

In the physical decomposition we decompose the timescale due to the physical contributions (e.g., conservation laws, time discretization methods), that are restricted by physical parameters (e.g., Courant-Friedrichs-Levy [CFL] condition), or very dominant physical parameters (e.g., reaction scales), see [94].

In the mathematical decomposition we underlie the operator equation and solve an eigenvalue problem with respect to the possible operators for the splitting method. We decide the splitting process taking the spectrum and the maximal eigenvalues into account. A second and more abstract idea is a decomposition involving only the spectrum of the operator, where we decouple the spectrum of the operator and compute new operators using the spectrum as a criterion for the timescales.

2.1.2 Physical Decomposition

For the physical decomposition method, the underlying physical parameters determine in which operators the equation is decoupled. Often the choice of the operators is obvious (e.g., for a problem in which the timescales of the physical behaviors are known); in this case, the direct method is possible.

2.1.2.1 Direct Decoupling Method

Often the values of the physical parameters are so obvious that we can determine the operators for the splitting methods.

The criteria for the direct decoupling method are strong anisotropy in the space dimensions, obvious timescales (e.g., fast reaction process and slow transport process, or different physical scales (e.g., growth and decay processes)).

For these obvious problems we can directly choose the form of the operators (e.g., the flow operator and reaction operator).

In an example for the direct decoupling method, we focus on the parabolic equation:

$$\frac{\partial c}{\partial t} = D_1(x,y)\frac{\partial^2 c}{\partial x^2} + D_2(x,y)\frac{\partial^2 c}{\partial y^2}, \text{ for } (x,y) \in \Omega, \ t \in [0,T], \quad (2.1)$$

$$c(x,y,0) = c_0(x,y), \text{ for } (x,y) \in \Omega, \quad (2.2)$$

$$c(x,y,t) = g(x,y,t), \text{ for } (x,y) \in \partial\Omega, \ t \in [0,T], \quad (2.3)$$

where the anisotropy of the heat operator is apparent:

$$\max_{(x,y)\in\Omega} D_1(x,y) << \min_{(x,y)\in\Omega} D_2(x,y).$$

Therefore, the decoupling in the spatial dimensions is possible, see [120] and [202].

2.1.2.2 Decoupling Method with Respect to Numerical Methods

The next decoupling possibility we discuss is after the discretization in time and space. Based on our discretization methods, stability conditions are necessary to obtain stable solutions, see [178].

Due to this restriction, the influence of the physical parameters is important for the stability criteria.

One such condition is the *Courant-Friedrichs-Lewy [CFL] condition*, see [51], in which no explicit, unconditionally stable, consistent finite difference schemes for hyperbolic initial value problems exist.

So the CFL condition gives the relation between the spatial and the time discretization with respect to the physical parameters, for which we have unconditionally stable results.

This condition can be used for decoupling the equations into different operators, when they present different timescales.

In the example for the decoupling method with respect to the numerical methods, we examine the transport-reaction equation:

$$\frac{\partial c}{\partial t} = v \cdot \nabla c + \lambda c, \text{ for } x \in \Omega, \ t \in [0,T], \quad (2.4)$$

$$c(x,0) = c_0(x), \text{ for } x \in \Omega, \quad (2.5)$$

$$c(x,t) = g(x,t), \text{ for } x \in \partial\Omega, \ t \in [0,T], \quad (2.6)$$

where the velocity parameter is given as $v = (v_1, \ldots, v_d)^T \in \mathbb{R}^{d,+}$, the reaction parameter is given as $\lambda \in \mathbb{R}^+$, and the spatial variables are $x = (x_1, \ldots, x_d)^T \in \Omega \subset \mathbb{R}^{d,+}$.

After the space and time discretization with first-order finite difference methods in time and space, we obtain the following CFL conditions for $d = 2$:

$$CFL_{\text{flow},x_1} = |\frac{v_{x_1}\tau}{\Delta x_1}| \leq 1, \quad (2.7)$$

$$CFL_{\text{flow},x_2} = |\frac{v_{x_2}\tau}{\Delta x_2}| \leq 1, \quad (2.8)$$

$$CFL_{\text{react}} = |\lambda\,\tau| \leq 1, \quad (2.9)$$

where τ is the time-step, Δx_1 and Δx_2 are the spatial steps.

With respect to the different scales, we determine the operators in the splitting methods, such that the similar CFL conditions are grouped together, for example,

$$CFL_{\text{flow},x_1} \approx CFL_{\text{react}} << CFL_{\text{flow},x_2}, \qquad (2.10)$$

so we select the operator A as the x_1-direction of the flow plus the reaction term and the operator B goes into the x_2-direction of the flow, see [145] and [179].

REMARK 2.1 Due to the discretization methods and their underlying stability criteria, there are further conditions for special evolution equations, such as the Neumann number for the diffusion equation and Prandtle number for the Navier-Stokes equation. These conditions present the physical behavior in the stability and can be used as a decoupling criterion, see [135], [179]. ☐

2.1.3 Mathematical Decomposition

In the mathematical decomposition method, we concentrate on the underlying operators, which means the scales of the operators, as the eigenvalues are indicators for the decomposition.

2.1.3.1 Decomposition with Respect to the Maximal Eigenvalues

In this method, we assume an ordinary differential equation with two different operators that are given by matrices.

We deal with the following equation:

$$\frac{dc(t)}{dt} = Ac(t) + Bc(t), \text{ for } t \in [0, T], \qquad (2.11)$$

$$c(0) = c_0, \qquad (2.12)$$

where $A, B : \mathbb{R}^m \to \mathbb{R}^m$ are example matrices. $c = (c_1, \ldots, c_m)^T$ is the solution vector.

DEFINITION 2.1 The spectral radius of a matrix M is defined by

$$\rho(M) = \max_{\lambda \in \sigma(M)} |\lambda|,$$

where $\sigma(M)$ is the spectrum of M and contains all eigenvalues.

We apply the spectral radius to our operators in (2.11) and have an approximation of the spectral radius, given as

$$\frac{||Ac_{\text{approx}}(t^{n+1})||}{||c_{\text{approx}}(t^{n+1})||} \approx \rho(A), \tag{2.13}$$

$$\frac{||Bc_{\text{approx}}(t^{n+1})||}{||c_{\text{approx}}(t^{n+1})||} \approx \rho(B), \tag{2.14}$$

where $|| \cdot ||$ is a vector norm on $\mathbb{R}^n$ and $c_{\text{approx}}(t^{n+1})$ is the numerical approximation of the Equation (2.11). The time-step $\tau_n = t^{n+1} - t^n$ is performed with an implicit method, with no restriction on τ_n.

REMARK 2.2 In the following we use the notation of a stiff operator, see [124]. Stiffness is not a mathematical definition, because no quantification is given for "large". We will instead use stiff operators, if the approximated spectral radius ρ is sufficiently large (e.g., relative to another spectral radius for example with a scale of 10^{-4}). □

Based on the approximated eigenvalues, we can decide the timescales for the splitting method, thus we have for the cases

1. $\rho(A) << \rho(B)$, the operator A has a large time-step with $\tau_{n,A} \approx \frac{1}{\rho(A)}$ and can be denoted as the nonstiff operator, where the operator B has a smaller time-step of $\tau_{n,B} \approx \frac{1}{\rho(B)}$ and can be denoted as the stiff operator.

2. $\rho(A) >> \rho(B)$, the operator A has a small time-step with $\tau_{n,A} \approx \frac{1}{\rho(A)}$ and can be denoted as the stiff operator, where the operator B has a large time-step of $\tau_{n,B} \approx \frac{1}{\rho(B)}$ and can be denoted as the nonstiff operator.

3. $\rho(A) \approx \rho(B)$, operators A and B have the same scales and there is no need to decouple the operators.

The operator scales may be defined in the following ways.

DEFINITION 2.2
We assume $\rho(A)$ to be the spectral radius of A and $\rho(B)$ to be the spectral radius of B.

Then we have the following definitions:

a) Two operators have the same scales if we have: $\rho(A) \approx \rho(B)$.

b) Two operators have large different scales if we have: $\rho(A) << \rho(B)$ (operator B is stiff, operator A is nonstiff) or $\rho(B) << \rho(A)$ (operator A is stiff, operator B is nonstiff).

We quantify the inequality as follows:
$$\rho(A) = c_{\text{large}} \, \rho(B), \text{ or } \rho(B) = c_{\text{large}} \, \rho(A),$$
where the factor $c_{\text{large}} \in [10^{-10}, 10^{-4}]$.

c) Two operators have moderately different scales if we have: $\rho(A) < \rho(B)$ (operator B is more stiff than operator A) or $\rho(B) < \rho(A)$ (operator A is more stiff than operator B).

We quantify the inequality as follows:
$$\rho(A) = c_{\text{small}} \, \rho(B), \text{ or } \rho(B) = c_{\text{small}} \, \rho(A),$$
where the factor $c_{\text{small}} \in [10^{-4}, 1]$.

For larger scales, for example $c_{\text{large}} \leq 10^{-10}$, we propose skipping the time discretization and treating the differential equation as an algebraic equation. In this case, we obtain DAEs (differential algebraic equations), see [153].

REMARK 2.3 To apply the operator-splitting method to a particular equation, we can deal with the stiff or the nonstiff scales and optimize our methods according to these time-steps. This gives more effective time-steps and saves computational time with respect to larger time-steps. ▯

2.1.3.2 Decomposition with Respect to Appropriate Norms

In this method, we assume the more interesting case of partial differential equation with two different operators that are given by a spatial discretization. We deal with the following equation:

$$\frac{dc(t)}{dt} = Ac(t) + Bc(t), \text{ for } t \in [0, T], \tag{2.15}$$

$$c(0) = c_0, \tag{2.16}$$

where A, B are unbounded operators of an analytic semigroup, and $c = (c_1, \ldots, c_m)^T$ is the solution vector.

Then we can found some norms $|| \cdot ||$ in which we can find some bounds with the operators.

We apply the appropriate norms to our operators in (2.19) given as

$$\frac{||Ac(t)||}{||c(t)||} \leq \lambda_1, \ t > 0, \tag{2.17}$$

$$\frac{||Bc(t)||}{||c(t)||} \leq \lambda_2, \ t > 0, \tag{2.18}$$

where $|| \cdot ||$ is an appropriate norm, and $\lambda_1, \lambda_2 \in \mathbb{R}^+$.

2.1.3.3 Abstract Decomposition Based on the Spectrum of the Operators (Symmetric Case)

A more abstract idea is to define a decoupling index for the spectrum of the underlying operator, which results after applying the spatial discretization of the partial differential equations. We assume to have only symmetric matrices in our operator equation for this case. We consider the following equation

$$\frac{dc(t)}{dt} = A_{\text{full}}\, c(t), \text{ for } t \in [0, T], \tag{2.19}$$

$$c(0) = c_0, \tag{2.20}$$

where $A_{\text{full}} : \mathbb{R}^m \to \mathbb{R}^m$ is, for example, a matrix that is derived from the spatial discretization of a parabolic differential equation. $c = (c_1, \ldots, c_m)^T$ is the solution vector, where we assume $m \in \mathbb{N}^+$ to be even.

The underlying spectrum of the operator A_{full} is given as

$$\sigma(A_{\text{full}}) = \{\lambda_1, \ldots, \lambda_m\}, \tag{2.21}$$

and we assume $\lambda_1 \leq \ldots \leq \lambda_m$ therefore, the eigenvalue problem is given as

$$X^T A_{\text{full}} X = \Lambda, \tag{2.22}$$

where Λ is the diagonal matrix with the eigenvalues $\lambda_1 < \lambda_2 < \ldots < \lambda_m$ and therefore the possible time-steps are $\tau_1 = \frac{1}{\lambda_1} > \ldots > \tau_m = \frac{1}{\lambda_m}$.

In the following we assume the same matrix X for diagonalizing the operator matrices A and B.

For the different timescales we decouple the spectrum in a large timescale $\tau_{m/2}$ and in a small timescale τ_m, so we define the new operators as

$$A = X \Lambda_A X^T, \tag{2.23}$$

where Λ_A is given with the eigenvalues $\lambda_1, \ldots, \lambda_{m/2}$, the other entries are zero, and

$$B = X \Lambda_B X^T, \tag{2.24}$$

where Λ_B is given with the eigenvalues $\lambda_{m/2+1}, \ldots, \lambda_m$, the other entries are zero.

Based on these operators, we apply the decomposition methods.

REMARK 2.4 For the application of the abstract decomposition the prestep method of computing the eigenvalues is necessary. Often the eigenvalue problem can be computed with the Lanczos method or an inverse iteration, see [181]. In our contributions, we deal with the physical decomposition, which is more applied. □

In the next section we introduce the numerical analysis for the decomposition methods.

2.1.4 Numerical Analysis of the Decomposition Methods

The underlying model equations for the multi-physics problems are evolution systems of partial differential equations. To solve such equations, numerical methods as discretization and solver methods have to be studied. We concentrate on the coupling of the equation systems and study decoupled equations with respect to large time-steps.

The numerical analysis for the partial differential equations is studied in [65], [200], and [201].

The large time-steps specialization for evolution equations is accomplished with mixed discretization or splitting methods. This specialization is studied in [124] for convection-diffusion equations, and in [150] for hyperbolic problems, in [118] for Hamiltonian problems. Decompositions of the equation systems in space and time are studied using domain decomposition in [179], [189], and [197]; time decomposition in [184] and [194]; and the classical splitting method in [154] and [185].

To obtain large time-steps of evolution equations, the application of implicit time discretization methods, as discussed in the numerical methods for ordinary differential equations, see [116] and [117], are proposed. Mixed methods of implicit and explicit discretization methods can also be applied as needed for decoupling stiff and nonstiff operators, as in references [171] and [172].

We further treat spatial adaptive methods, which refine the domain regions, and apply standard implicit discretization methods, see, for example, [10], [28], [29], [39], and [64]. The spatial error estimates decrease the local errors; therefore, the stiffness of the underlying operators decreases, too, see [200] and [201].

2.1.5 Decoupling Index

The benefit, of the splitting methods were fine in the 1960s, when computer power was limited to one-dimensional problems and splitting was done with respect to decoupling into one-dimensional problems. One of the first studies can be found in [185], and [154] describes the decoupling with respect to higher-order ideas.

Nowadays faster computers and fast solver methods exist (e.g., fast-block ILU solvers [196], multigrid methods [112] and [115]), and such methods cannot become more efficient due to the complexity of the underlying equation systems. So the splitting methods can be addressed in the context of accelerating a method (e.g., with Newton method [173], a solver method, or by decoupling multiscales for special equations), see [168].

Therefore, it is important to define an index to classify the decomposition. One idea is to define the decoupling index as a spectrum of the different operators of the equation, see Section 2.1.3.

In our applications, we deal with the more physically oriented decomposition that allows us to decouple the equation on the level of the partial differential equation.

In the next section we deal with the discretization.

2.2   Discretization

In discretization, we deal with the underlying methods to obtain a system of ordinary differential equations.

Here, the underlying idea is to use spatial discretization methods to transform our partial differential equations into ordinary differential equations. The resulting operator equations can be treated with the decomposition methods, and the analysis for consistency and stability is based on the semigroup theory, see [200] and [201].

We concentrate on the Galerkin method to perform the spatial discretization.

2.2.1   Galerkin Method

With the Galerkin method, we obtain approximates, the *Galerkin solutions*, and reach an operator equation that can be treated as a system of ordinary differential equations.

The following points describe the Galerkin method:

- Multiplication of the differential equation for u by the functions $v \in K$ and integration over Ω with subsequent integration by parts yields the generalized problem.

- The function class K is chosen such that a sufficient smooth solution u of the generalized problem is also a solution of the original classical problem when the data are sufficiently smooth (boundary, coefficients, boundary and initial value, etc.).

- Restriction to u and v in the generalized problem to appropriate finite-dimensional subspaces.

The classical derivatives are replaced by generalized (weak) derivatives. The realization of the Galerkin method is accomplished with the basis functions $w_1, \ldots, w_m$, see [200].

In the following, we realize the abstract ideas to parabolic and hyperbolic differential equations, with the aim of deriving the operator equation. With this abstract generalized problem of coupled systems, we can apply the splitting methods.

2.2.2 Parabolic Differential Equation Applied with the Galerkin Method

We now apply the Galerkin method to a parabolic differential equation given as

$$\frac{du(x,t)}{dt} - \Delta u(x,t) = f(x,t), \text{ in } \Omega \times [t_0, T], \tag{2.25}$$

$$u(x,t) = 0, \text{ on } \partial\Omega \times [t_0, T] \text{ (boundary condition)},$$

$$u(x,0) = u_0(x), \text{ on } \Omega \text{ (initial condition)},$$

where $\Omega \in \mathbb{R}^{N,+}$ is a Lipschitz domain, N is the spatial dimension, and $[t_0, T] \subset \mathbb{R}^+$ is the temporal domain.

The generalized problem of the equation (B.1) yields to the following formulation:

We seek a function $u \in C([t_0, T]; L^2(\Omega)) \cap C((t_0, T]; H_0^1(\Omega))$, such that

$$\frac{d}{dt} \int_\Omega u(x,t)v(x) \, dx + \int_\Omega \left(\sum_{i=1}^N D_i u(x,t) \, D_i v(x) - f \, v(x) \right) dx = 0,$$

$$\text{in } \in [t_0, T], \tag{2.26}$$

$$u(x,t) = 0 , \text{ on } \partial\Omega \times [t_0, T] \text{ (boundary condition)},$$

$$u(x,0) = u_0(x), \text{ on } \Omega \text{ (initial condition)},$$

holds for all $v \in C_0^\infty(\Omega) \cap H_0^1(\Omega)$. Further, we assume $f(x,t) \in L^2([t_0, T]; L^2(\Omega))$ and $u_0(x) \in H_0^1(\Omega)$.

The function $v(x)$ depends only on the spatial variable x.

In order to construct an approximate solution u_m with the Galerkin method, we make the following attempt:

$$u_m(x,t) = \sum_{k=1}^m c_{km}(t) \, w_k(x), \tag{2.27}$$

where the unknown coefficients $c_{k,m}$ depend on time. We replace u in the generalized problem (B.2) by u_m and require that (B.2) holds for all $v \in span\{w_1, \ldots, w_m\}$. Then we obtain the Galerkin equations for $j = 1, \ldots, m$:

$$\sum_{k=1}^m c'_{km}(t) \int_\Omega w_k \, w_j \, dx + \sum_{k=1}^m c_{km}(t) \int_\Omega \sum_{i=1}^N D_i w_k \, D_i w_j \, dx$$

$$= \int_\Omega f \, w_j \, dx. \tag{2.28}$$

Here we have a linear system of first-order ordinary differential equations for the real functions $c_{1m}, \ldots, c_{mm}$. The number of the dimension for the test space is given as m, and for $m \to \infty$ we obtain the convergence of u_m to u.

Based on this notation, we can write in abstract operator equations:

$$Mc'_m(t) + Ac_m(t) = F(t) , \ t \in [t_0, T], \tag{2.29}$$

$$c_m(0) = \alpha_m, \tag{2.30}$$

where $c_m = (c_{1m}, \dots, c_{mm})^T$ is the solution vector and $\alpha_m = (\alpha_{1m}, \dots, \alpha_{mm})^T$ is the initial condition, and

$$M = \begin{pmatrix} \int_\Omega w_1 \, w_1 \, dx & \cdots & \int_\Omega w_1 \, w_m \, dx \\ \vdots & \ddots & \vdots \\ \int_\Omega w_m \, w_1 \, dx & \cdots & \int_\Omega w_m \, w_m \, dx \end{pmatrix}, \tag{2.31}$$

$$A = \begin{pmatrix} \int_\Omega \sum_{i=1}^N D_i w_1 \, D_i w_1 \, dx & \cdots & \int_\Omega \sum_{i=1}^N D_i w_1 \, D_i w_m \, dx \\ \vdots & \ddots & \vdots \\ \int_\Omega \sum_{i=1}^N D_i w_m \, D_i w_1 \, dx & \cdots & \int_\Omega \sum_{i=1}^N D_i w_m \, D_i w_m \, dx \end{pmatrix}, \tag{2.32}$$

and $F(t) = (\int_\Omega f \, w_1 dx, \dots, \int_\Omega f \, w_m dx)^T$ are the operators.

The initial conditions $c_m(0)$ correspond to the approximately initial conditions of the original problem, so we choose a sequence

$$u_{m,0}(x) = \sum_{k=1}^m \alpha_{km} w_k(x), \tag{2.33}$$

where $u_{m,0}$ converges to u_0 as $m \to \infty$.

For the boundary conditions we also have

$$u_{m,0}(x,t) = 0 \text{ on } \partial\Omega \times [t_0, T]. \tag{2.34}$$

REMARK 2.5 If we replace the linear operators by nonlinear operators in (B.1), we obtain a nonlinear system of ordinary differential equations. This can be treated by linearization and we obtain again our linear ordinary differential equations. □

REMARK 2.6 Extensions with convection and reaction terms can also be performed with the parabolic differential equation. The generalization is shown in references [81], [124]. Because of the operator notation as a further simplification of the presented partial differential equations, we concentrate on the diffusion term. □

2.2.3 Hyperbolic Differential Equation Applied with the Galerkin Method

In this section, we apply the Galerkin method on a hyperbolic differential equation, given as

$$\frac{d^2 u(x,t)}{dt^2} - \Delta u(x,t) = f(x,t), \text{ in } \Omega \times [t_0, T], \tag{2.35}$$

$$u(x,t) = 0, \text{ on } \partial\Omega \times [t_0, T] \text{ (boundary condition)},$$

$$u(x,0) = u_0(x), \text{ on } \Omega \text{ (initial condition)},$$

$$u_t(x,0) = u_1(x), \text{ on } \Omega \text{ (initial condition)},$$

where f, u_0, and u_1 are given and we seek u.

A generalized approach to equation (B.11) yields to the formulation presented below.

We seek a function $u(x,t) \in C^1([t_0,T]; L^2(\Omega)) \cap C([t_0,T]; H_0^1(\Omega))$, such that the following

$$\frac{d^2}{dt^2} \int_\Omega u(x,t) v(x) \, dx$$

$$+ \int_\Omega \left(\sum_{i=1}^N D_i u(x,t) \, D_i v(x) - f(x,t) \, v(x) \right) dx = 0,$$

$$\text{in } \Omega \times [t_0, T], \tag{2.36}$$

$$u(x,t) = 0, \text{ on } \partial\Omega \times [t_0, T] \text{ (boundary condition)},$$

$$u(x,0) = u_0(x), \text{ on } \Omega \text{ (initial condition)},$$

$$u_t(x,0) = u_1(x), \text{ on } \Omega \text{ (initial condition)},$$

holds for all $v(x) \in C_0^\infty(\Omega) \cap H_0^1(\Omega)$. We also assume $f(x,t) \in L^2([t_0,T]; L^2(\Omega))$, $u_0(x) \in H_0^1(\Omega)$ and $u_1(x) \in L^2(\Omega)$.

The function $v(x)$ depends only on the spatial variable x.

The following is our attempt to construct an approximate solution u_m with the Galerkin method, thus making the following attempt:

$$u_m(x,t) = \sum_{k=1}^m c_{k,m}(t) \, w_k(x), \tag{2.37}$$

where the unknown coefficients $c_{k,m}$ are time dependent. We replace u in the generalized equation (B.12) by u_m and require that (B.12) holds for all $v \in \text{span}\{w_1, \ldots, w_m\}$. Then we obtain the Galerkin equations for $j = 1, \ldots, m$:

$$\sum_{k=1}^m c''_{km}(t) \int_\Omega w_k \, w_j \, dx + \sum_{k=1}^m c_{km} \int_\Omega \sum_{i=1}^N D_i w_k \, D_i w_j \, dx$$

$$= \int_\Omega f \, w_j \, dx, \tag{2.38}$$

where we have a linear system of second-order ordinary differential equations for the real functions $c_{1m}, \ldots, c_{mm}$.

Based on this notation, we can write in abstract operator equations:

$$Mc_m''(t) + Ac_m(t) = F(t), \ t \in [t_0, T], \tag{2.39}$$

$$c_m(0) = \alpha_m, \tag{2.40}$$

$$c_m'(0) = \beta_m, \tag{2.41}$$

where c_m is the solution vector; α_m, β_m are the initial conditions; and M and A are given as in Equations (B.7) and (B.8).

The initial conditions $c_m(0)$ and $c_m'(0)$ correspond approximately to the initial conditions of the original problem, so we choose a sequence

$$u_{m,0}(x) = \sum_{k=1}^{m} \alpha_{km} w_k(x), \tag{2.42}$$

$$u_{m,1}(x) = \sum_{k=1}^{m} \beta_{km} w_k(x), \tag{2.43}$$

where $u_{m,0}$ converges to u_0 and $u_{m,1}$ to u_1 as $m \to \infty$.

For the boundary conditions, we also have

$$u_{m,0}(x,t) = 0 \text{ on } \partial\Omega \times [t_0, T]. \tag{2.44}$$

REMARK 2.7 If we replace the linear operators by nonlinear operators in (B.11), we obtain a nonlinear system of ordinary differential equations. This can be treated by linearization (e.g., Newton method or fixed-point iterations), see [131], and we obtain again our linear ordinary differential equations.
☐

2.2.4 Operator Equation

For a more abstract treatment of our splitting methods, see Chapter 3 where we discuss this type of operator equation:

$$A_1 \frac{d^2}{dt^2} C_1(u(t), t) + A_2 \frac{d}{dt} C_2(u(t), t) + B(u(t), t) = 0, \ \forall t \in [t_0, T], \tag{2.45}$$

where A_1, A_2, B, C_1, C_2 are positive definite and symmetric operators, thus at a minimum, they are also boundable operators for all $t \in [t_0, T]$.

We also have to reset the notation c_m to u for the abstract treatment. This can be done by applying the test equations, see equation (B.13).

For the operator equations, we can, in Chapter 3, treat our underlying splitting methods with the semigroup theory.

2.2.5 Semigroup Theory

With the concept of a semigroup, we can describe time-dependent processes in nature in terms of the functional analysis. We describe an introduction that is needed in the further sections. An overview of the semigroup theory can be found in [12], [13], [65], [191], [198], [200], and [201]. The key relations are

$$S(t + s) = S(t)S(s), \quad \forall t, s \in \mathbb{R}^+, \tag{2.46}$$
$$S(0) = I, \tag{2.47}$$

and we have the following definition of a generator of a semigroup.

DEFINITION 2.3 A semigroup $\{S(t)\}$ on a Banach space X consists of a family of operators $S(t) : X \to X$ for all $t \in \mathbb{R}^+$ with (B.22) and (B.23). The generator $B : D(B) \subset X \to X$ of the semigroup $\{S(t)\}$ is defined by

$$Bw = \lim_{t \to 0^+} \frac{S(t)w - w}{t}, \tag{2.48}$$

where w belongs to $D(B)$ (and $D(B)$ is the domain of B) if the limit of (B.24) exists.

A one-parameter group $\{S(t)\}$ on the Banach space X consists of a family of operators $S(t) : X \to X$ for all $t \in \mathbb{R}$; with (B.22) for all $t, s \in \mathbb{R}$; and (B.23) as the initial condition.

REMARK 2.8 A family $\{S(t)\}$ of linear continuous operators from X to itself, which satisfies the conditions (B.22) and (B.23), is called a *continuous semigroup* of linear operators or simply a C_0 *semigroup*. ▯

2.2.6 Classification of Semigroups

Let $\mathcal{S} = \{S(t)\}$ be a semigroup of the Banach space X. The following list gives the classification of the semigroups according to their characteristics:

1. $\mathcal{S}$ is called strongly continuous, if for all $S(t)$ holds, that $t \mapsto S(t)w$ is continuous on $\mathbb{R}^+$ for all $w \in X$ that is,

$$\lim_{t \to s} S(t)w = S(s)w, \quad \forall s \in \mathbb{R}^+. \tag{2.49}$$

2. $\mathcal{S}$ is called uniformly continuous, if all operators $S(t) : X \to X$ are linear and continuous, and $t \mapsto S(t)$ is continuous on $\mathbb{R}^+$ with respect to the operator norm that is,

$$\lim_{t \to s} ||S(t) - S(s)|| = 0, \quad \forall s \in \mathbb{R}^+. \tag{2.50}$$

3. S is called nonexpansive, if all operators $S(t) : X \to X$ are nonexpansive and

$$\lim_{t \to 0^+} S(t)w = w, \ \forall w \in X. \tag{2.51}$$

4. S is called a linear semigroup, if all operators $S(t) : X \to X$ are linear and continuous.

5. S is called an analytical semigroup, if we have an open sector Σ^α and a family of linear continuous operators $S(t) : X \to X$ for all $t \in \Sigma^\alpha$ with $S(0) = I$ and the following properties:
   (a) $t \mapsto S(t)$ is an analytical map from Σ^α into $L(X,Y)$.
   (b) $S(t + s) = S(t)S(s)$ for all $t, s \in \Sigma^\alpha$.
   (c) $\lim_{t \to 0^+} S(t)w = w$ in Σ^α for all $w \in X$.
   where an open sector is defined as:

$$\Sigma^\alpha = \{z \in \mathbb{C} : -\alpha < \arg(z) < \alpha , \ z \neq 0\}, \tag{2.52}$$

   see also Figure B.1.

6. S is called a **bounded analytical semigroup** if we have an analytical semigroup and additionally the property for each $\beta \in]0, \alpha[$ is satisfied:

$$\sup_{t \in \Sigma_\beta^\alpha} \|S(t)\| < \infty, \tag{2.53}$$

where $\Sigma_\beta^\alpha = \{z \in \mathbb{C} : -\beta < \arg z < \beta , \ z \neq 0\}$.

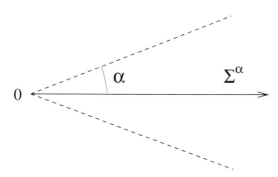

FIGURE 2.1: Sector for the analytical semigroup.

We give examples for the different classifications of the semigroups in the following parts.

Example 2.1
(i) If $B : D(B) \subset X \to X$ is a linear self-adjoint operator (i.e., (Bu, v) = (u, Bv), see [200]), on the Hilbert space X with $(Bu, u) \leq 0$ on $D(B)$, then B is the generator of a linear nonexpansive semigroup. In particular, such semigroups can be used to describe heat conduction and diffusion processes. In terms of the general functional calculus for self-adjoint operators, this semigroup is given by $\{\exp(tB)\}$.

(ii) If $H : D(H) \subset X \to X$ is a linear self-adjoint operator on the complex Hilbert space X, then $-iH$ generates a one-parameter unitary group. Such groups describe the dynamics of quantum systems. The operator H corresponds with the energy of the quantum system and is called the Hamiltonian of the system. In terms of the general functional calculus for self-adjoint operators on the complex space, this semigroup is given by $\{\exp(-itH)\}$.

(iii) If $C : D(C) \subset X \to X$ is a skew-adjoint operator (i.e., $(Cu, v) = (u, -Cv)$, see [200]) on the real Hilbert space X, then C is the generator of a one-parameter unitary group. Such semigroups describe, for example, the dynamics of wave processes. ☐

In our monograph we will discuss examples (i) and (iii) (i.e., the self-adjoint and the skew-adjoint operator on the real Hilbert space X).

In addition, for realistic application to heat equations, we must assume unbounded operators because of the irreversibility of the processes (for more, see [200]).

2.2.7 Abstract Linear Parabolic Equations

For a discussion of parabolic equations, we consider the equations in a notation of an abstract initial value problem, given as

$$u'(t) = Bu(t) + f(t), \text{ for } t_0 < t < T, \tag{2.54}$$
$$u(0) = u_0,$$

and the solution of (B.30):

$$u(t) = S(t - t_0) u_0 + \int_{t_0}^{t} S(t - s) f(s) \, ds, \tag{2.55}$$

where the integration term is a convolution integral, see [200], and can be solved numerically with Runge-Kutta methods, see [116] and [117].
We also have the following assumptions:

Assumption 2.1 *(H1) Let $\{S(t)\}$ be a strongly continuous linear semigroup on the Banach space X over $\mathbb{R}$ or $\mathbb{C}$ with the generator B that is, $\{S(t)\}$*

is a semigroup of linear continuous operators $S(t) : X \to X$ for all $t \geq 0$, and $t \mapsto S(t)w$ is continuous on $\mathbb{R}^+$ for all $w \in X$.

(H2) The function $f : [t_0, T[\to X$ is continuous.

THEOREM 2.1

Assuming $(H1)$ and $(H2)$, it holds that:

(a) *There exists at most one classical solution of (B.30), and each classical solution is also a mild solution (weak solution).*

(b) *If $f \in C^1$ and $w \in D(B)$, then the mild solution (B.31) is also a classical solution of (B.30).*

(c) *If the operator $B : X \to X$ is linear and continuous, then, for each $w \in S$ and each continuous f, the mild solution (B.31) is also a classical solution of (B.30).*

This illustrates the importance of the semigroups for the solution of the initial value problem (B.30).

COROLLARY 2.1

(i) There exist constants $C \geq 1$ and $\alpha \geq 0$, such that

$$||S(t)|| \leq C \exp(\alpha t), \ \forall \, t \geq 0. \tag{2.56}$$

(ii) The generator $B : D(B) \subset X \to X$ of the semigroup $\{S(t)\}$ is a linear graph-closed operator and $D(B)$ is dense in X.
(iii) The semigroup is uniquely determined by its generator.

With this notation of the semigroup for the parabolic equation, we can abstractly treat the splitting methods, see Chapter 3.

2.2.8 Abstract Linear Hyperbolic Equations

For the discussion about the hyperbolic equations, we consider the equations in a notation of an abstract initial value problem given as

$$u''(t) + Au(t) = f(u(t)), \text{ for } 0 < t < \infty, \tag{2.57}$$
$$u(0) = u_0, \ u'(0) = u_1.$$

We also have the following assumptions:

Assumption 2.2 *(H1) The linear operator $A : D(A) \subset X \to X$ is self-adjoint and strongly monotone on the Hilbert space over $\mathbb{R}$ or $\mathbb{C}$. Let X_E be the energetic space of A with the norm $|| \cdot ||_E$ that is, X_E is the* completion of $D(A)$ with respect to the energetic scalar product $(u, v)_E = (Au, v)$. In other words, A is a symmetric and positive definite operator, see also [65].

(H2) *The operator $f : X_E \to X$ is locally Lipschitz continuous that is, for each $R > 0$ there is a constant L such that*

$$\|f(u) - f(v)\| \leq L\|u - v\|_E, \tag{2.58}$$

for all $u, v \in X_E$ with $\|u\|_E, \|v\|_E \leq R$.

Setting $v = u'$, we rewrite Equation (B.33) into a first-order system and achieve

$$\begin{pmatrix} u' \\ v' \end{pmatrix} = \begin{pmatrix} 0 & I \\ -A & 0 \end{pmatrix} \begin{pmatrix} u \\ v \end{pmatrix} + \begin{pmatrix} 0 \\ f \end{pmatrix}. \tag{2.59}$$

Setting $z = (u, v)$ and rewriting (B.34), we obtain

$$z''(t) = Cz(t) + F(z(t)), \text{ for } 0 < t < \infty, \tag{2.60}$$
$$z(0) = z_0.$$

Let $Z = X_E \times X$ and $D(C) = D(A) \times X_E$.

If we use the assumption $(H1)$, then the operator C is skew-adjoint and generates a one-parameter unitary group $\{S(t)\}$.

The applications to this semigroup are discussed in Chapter 4. We discuss the iterative splitting method with respect to consistency and stability analysis in the semigroup notation.

For many applications, nonlinear semigroups are important. Therefore, in the next subsection, we describe the notations and important results for the abstract nonlinear semigroup theorem, which we will need in the upcoming chapters.

2.2.9 Nonlinear Equations

In this section, we discuss the abstract semigroup theory for nonlinear operators by introducing certain nonlinear semigroups, generated by convex functions, see [65] and [166]. These can be applied for various nonlinear second-order parabolic partial differential equations.

We apply the nonlinear semigroups to our nonlinear differential equations.

In the following, we begin with a Hilbert space H and take $I : H \to (-\infty, +\infty]$ to be convex, proper, and lower semicontinuous.

For simplicity, we also assume that ∂I is densely defined that is, $\overline{D(\partial I)} = H$.

We further propose to study the nonlinear differential equation given as

$$u'(t) + A(u(t)) \ni 0, \text{ for } 0 \leq t < \infty, \tag{2.61}$$
$$u(0) = u_0,$$

where $u_0 \in H$ is given and $A = \partial I$ is a nonlinear, discontinuous operator, and is also multivalued.

For the convex analysis, we also assume that (B.36) has a unique solution for each initial point u_0. We then write

$$u(t) = S(t)u_0, \text{ for } 0 < t < \infty, \tag{2.62}$$

and regard $S(t)$ as a mapping from H into H for each time point $t \geq 0$. We note that the mapping $u_0 \mapsto S(t)u_0$ is generally nonlinear.

As we defined the linear semigroup, we also have the following conditions, see [65]:

$$S(0)u_0 = u_0, \text{ for } u_0 \in H, \tag{2.63}$$

$$S(t+s)u_0 = S(t)S(s)u_0, \ t, s \geq 0, \ u_0 \in H, \tag{2.64}$$

the mapping $t \mapsto S(t)u_0$ is continuous from $[0, \infty)$ into H.

Then we arrive at a definition for the nonlinear semigroup.

DEFINITION 2.4

(i) A family $\{S(t)\}$ of nonlinear operator mappings H into H is called a nonlinear semigroup, if the conditions (B.38) and (B.39) are satisfied.

(ii) We say $\{S(t)\}$ is a contractive semigroup, if in addition there holds:

$$||S(t)u - S(t)\hat{u}|| \leq ||u - \hat{u}||, \ t \geq 0, u, \hat{u} \in H. \tag{2.65}$$

We could show that the operator $A = \partial I$ generates a nonlinear semigroup of contractions on H, so we could solve the ordinary differential equation:

$$u'(t) \in -\partial I(u(t)), \ t \geq 0, \tag{2.66}$$

$$u(0) = u_0, \tag{2.67}$$

which is well posed for a given initial point $u_0 \in H$. It is obtained by a type of infinite dimensional gradient flow, see [65].

For the theory, the idea in our monograph is to regularize or smooth the operator $A = \partial I$ with linearization (e.g., with Taylor expansions or fixed-point iterations with previous iterations), or by finding an A_λ with a resolvent as a regularization.

REMARK 2.9 For the regularization, the nonlinear resolvent J_λ can be defined as

DEFINITION 2.5 (1) For each $\lambda > 0$, we define the nonlinear resolvent $J_\lambda : H \to D(\partial I)$ by setting

$$J_\lambda[w] := u,$$

where u is a unique solution of

$$w \in u + \lambda \partial I[u].$$

(2) For each $\lambda > 0$ we define the Yoshida approximation $A_\lambda : H \to H$ by

$$A_\lambda[w] := \frac{w - J_\lambda[w]}{\lambda}, \quad w \in H. \tag{2.68}$$

Therefore, A_λ is a type of regularization or smoothing of the operator $A = \partial I$. ⧠

For the semigroup theory, we can define the resolvent with the infinitesimal generator A, see [198].

THEOREM 2.2
If $\lambda > 0$, then the operator $(\lambda I - A)$ admits an inverse
$$R(\lambda, A) = (\lambda I - A)^{-1} \in \mathcal{L}(X), \text{ and}$$

$$R(\lambda, A)x = \int_0^\infty \exp(-\lambda s) \, T_s x \, ds \text{ for } x \in X, \tag{2.69}$$

where T_s is the linear operator of the semigroups.
Thus, positive real numbers belong to the resolvent set $\rho(A)$ of A and we have a type of eigenvalue for the inverse operator.

These can be used to study the nonlinear problems with the help of the eigenvalue problems.

REMARK 2.10 The nonlinear semigroup theory can be applied for the analysis of nonlinear operator equations. For our consistency and stability analysis, we propose a linearization of the nonlinear operators and consider more the linear semigroup theory. ⧠

Chapter 3

Time-Decomposition Methods for Parabolic Equations

In this chapter we focus on the methods for decoupling multiphysical and multidimensional equations. The main idea is to decouple a complex equation in various simpler equations and to solve the simpler equations with adapted discretization and solver methods.

The methods are described in the literature for the basic studies in [193] and [185] and for an overview in [108].

In many applications in the past, a mixing of the various terms in the equations for the discretization and solver methods made it difficult to solve them together. With respect to the adapted methods for a simpler equation, the methods allow improved results for simpler parts.

In general, the simpler parts are collected and via the initial conditions the results are coupled together.

For notifications we distinguish between the splitting of the dimensions or the splitting of operators, named as dimensional-splitting or operator-splitting, respectively.

These flexibilities are used for the different operations.

The first splitting methods were developed in the 1960s or 1970s and were based on fundamental results of finite difference methods. A renewal of the methods was done in the 1980s while using the methods for complex processes underlying partial differential methods, cf. [52].

In particular for the more complex models in the enviromential-physics (e.g., contaminant transport in gas, fluid or porous media, cf. [202]), new higher-order methods were developed using the ideas of the ordinary differential equations. With these new methods, an error analysis for the splitting methods was developed. The theoretical background is described in the theorem of Baker-Campbell-Haussdorff, cf. [148] and [191], using the operator theory.

In our work we apply the splitting methods with respect to mixed methods based on the discretization methods. The main results for the decoupling of systems of mixed hyperbolic and parabolic equations are introduced. Our results for the mixed equations for convection-diffusion-reaction equations are based on the results described in the literature [81] and [193].

The best fit for the splitting methods is discussed with respect to the physical characteristics (e.g., Peclet-number, Prandtle-number, etc.). A new technique for physical-splitting methods is explained and the results are presented.

Also interesting iterative methods are introduced to fulfill the exactness of the splitting-method of higher-order and the approximative range of iterative methods, in which more effectivity is given.

In the next section we introduce the splitting methods.

The techniques for the applications for different operators are given as shown in Figure 3.1.

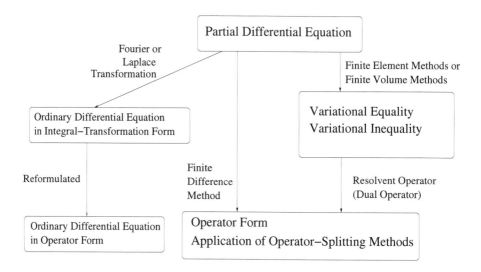

FIGURE 3.1: Application of the time-decomposition methods.

3.1 Introduction for the Splitting Methods

The natural way of decoupling an ordinary differential equation in simpler parts is done in the following way:

$$c'(t) = A\, c(t), \tag{3.1}$$

$$c'(t) = (A_1 + A_2)\, c(t), \tag{3.2}$$

where the initial conditions are $c^n = c(t^n)$. The operator A can be decoupled in the operators A_1 and A_2, cf. the introduction in [184].

Based on these linear operators, Equation 3.1 can be solved exactly. The solution is given as

$$c(t^{n+1}) = \exp(\tau A)\, c(t^n), \tag{3.3}$$

where the time-step is $\tau = t^{n+1} - t^n$.

The simplest operator-splitting methods are the sequential operator-splitting methods that are decoupling into two or more equations. One could analyze the error for the linear case by the Taylor expansion.

We deal with the following linear and sequential operator-splitting methods.

3.1.1   Classical Operator-Splitting Methods

In the following, we describe those traditional operator-splitting methods which are widely used for the solution to the real-life problems. We focus our attention on the case of two linear operators (i.e., we consider the Cauchy problem):

$$\partial_t c(t) = Ac(t) + Bc(t), \quad t \in (0,T), \quad c(0) = c_0, \tag{3.4}$$

whereby the initial function c_0 is given, and A and B are assumed to be bounded linear operators in the Banach-space $\mathbf{X}$ with $A, B : \mathbf{X} \to \mathbf{X}$. In realistic applications the operators correspond to physical operators (e.g., convection and diffusion operator).

3.1.2   Sequential Operator-Splitting Method

First, we describe the simplest operator-splitting, which is called *sequential operator-splitting*. The sequential operator-splitting method is introduced as a method, which solves two subproblems sequentially on subintervals $[t^n, t^{n+1}]$, where $n = 0, 1, \ldots, N-1$, $t^0 = 0$, and $t^N = T$. The different subproblems are connected via the initial conditions. This means that we replace the original problem (3.4) with the subproblems on the subintervals:

$$\frac{\partial c^*(t)}{\partial t} = Ac^*(t), \quad t \in (t^n, t^{n+1}) \quad \text{with } c^*(t^n) = c_{\text{sp}}^n, \tag{3.5}$$

$$\frac{\partial c^{**}(t)}{\partial t} = Bc^{**}(t), \quad t \in (t^n, t^{n+1}) \quad \text{with } c^{**}(t^n) = c^*(t^{n+1}),$$

for $n = 0, 1, \ldots, N-1$, whereby $c_{\text{sp}}^0 = c_0$ is given from (3.4). The approximated split solution at the point $t = t^{n+1}$ is defined as $c_{\text{sp}}^{n+1} = c^{**}(t^{n+1})$.

Clearly, the change of the original problems with the subproblems usually results in some error, called *local splitting error*. The local splitting error of the sequential operator-splitting method can be derived as follows:

$$\rho_n = \frac{1}{\tau_n} \left(\exp(\tau_n(A+B)) - \exp(\tau_n B)\exp(\tau_n A) \right) c_{\text{sp}}^n$$

$$= \frac{1}{2}\tau_n [A, B]\, c(t^n) + \mathcal{O}(\tau_n^2), \tag{3.6}$$

whereby the splitting time-step is defined as $\tau_n = t^{n+1} - t^n$. We define $[A, B] := AB - BA$ as the commutator of A and B. Consequently, the splitting error is $O(\tau_n)$ when the operators A and B do not commute. When the operators commute, then the method is exact. Hence, by definition, the sequential operator-splitting is called the *first-order splitting method.*

3.1.3 Symmetrically Weighted Sequential Operator-Splitting

For noncommuting operators, the sequential operator-splitting is not symmetric w.r.t. the operators A and B, and it has first-order accuracy. However, in many practical cases we require splittings of higher-order accuracy. We can achieve this by the following modified splitting method, called symmetrically weighted sequential operator-splitting, which is already symmetrical w.r.t. the operators.

The algorithms read as follows. We consider again the Cauchy problem (3.4), and we define the operator-splitting on the time interval $[t^n, t^{n+1}]$ (where $t^{n+1} = t^n + \tau_n$) as

$$\frac{\partial c^*(t)}{\partial t} = Ac^*(t), \quad \text{with } c^*(t^n) = c_{\text{sp}}^n, \tag{3.7}$$

$$\frac{\partial c^{**}(t)}{\partial t} = Bc^{**}(t), \quad \text{with } c^{**}(t^n) = c^*(t^{n+1}),$$

and

$$\frac{\partial v^*(t)}{\partial t} = Bv^*(t), \quad \text{with } v^*(t^n) = c_{\text{sp}}^n, \tag{3.8}$$

$$\frac{\partial v^{**}(t)}{\partial t} = Av^{**}(t), \quad \text{with } v^{**}(t^n) = v^*(t^{n+1}),$$

where c_{sp}^n is known.
Then the approximation at the next time level t^{n+1} is defined as

$$c_{\text{sp}}^{n+1} = \frac{c^{**}(t^{n+1}) + v^{**}(t^{n+1})}{2}. \tag{3.9}$$

The splitting error of this operator-splitting method is derived as follows:

$$\rho_n = \frac{1}{\tau_n}\{\exp(\tau_n(A + B)) - $$
$$-\frac{1}{2}[\exp(\tau_n B)\exp(\tau_n A) + \exp(\tau_n A)\exp(\tau_n B)]\}c(t^n). \tag{3.10}$$

An easy computation shows that in the general case,

$$\rho_n = O(\tau_n^2), \tag{3.11}$$

that is, the method is of second-order accuracy. We note that in the case of commuting operators A and B the method is exact that is, the splitting error vanishes.

3.1.4 Strang-Marchuk Operator-Splitting Method

One of the most popular and widely used operator splittings is the *Strang operator-splitting method (or Strang-Marchuk operator-splitting method)*, which reads as follows [185]:

$$\frac{\partial c^*(t)}{\partial t} = Ac^*(t), \text{ with } t^n \le t \le t^{n+1/2} \text{ and } c^*(t^n) = c_{\text{sp}}^n, \tag{3.12}$$

$$\frac{\partial c^{**}(t)}{\partial t} = Bc^{**}(t), \text{ with } t^n \le t \le t^{n+1} \text{ and } c^{**}(t^n) = c^*(t^{n+1/2}),$$

$$\frac{\partial c^{***}(t)}{\partial t} = Ac^{***}(t), \text{ with } t^{n+1/2} \le t \le t^{n+1} \text{ and } c^{***}(t^{n+1/2}) = c^{**}(t^{n+1}),$$

where $t^{n+1/2} = t^n + 0.5\tau_n$, and the approximation on the next time level t^{n+1} is defined as $c_{\text{sp}}^{n+1} = c^{***}(t^{n+1})$.

The splitting error of the Strang splitting is

$$\rho_n = \frac{1}{24}\tau_n^2([B, [B, A]] - 2[A, [A, B]]) \, c(t^n) + O(\tau_n^3), \tag{3.13}$$

see, for example, [185]. This means that this operator splitting is of second-order, too. We note that under some special conditions for the operators A and B, the Strang splitting has third-order accuracy and even can be exact, see [185].

In the next section, we present some other types of operator-splitting methods that are based on the combination of the operator-splitting and the iterative methods.

3.1.5 Higher-Order Splitting Method

The higher-order operator splitting methods are used for more accurate computations, but also with respect to more computational steps. These methods are often performed in quantum dynamics to approximate the evolution operator $\exp(\tau(A + B))$, see [42].

An analytical construction of higher-order splitting methods can be performed with the help of the BCH formula (Baker-Campbell-Hausdorff), see [118] and [199].

The reconstruction process is based on the following product of exponential functions:

$$\exp(\tau(A + B)) = \Pi_{i=1}^m \exp(c_i\tau A)\exp(d_i\tau B) + \mathcal{O}(\tau^{m+1}) = S(\tau), \tag{3.14}$$

where A, B are noncommutative operators, τ is the equidistant time-step, and $(c_1, c_2, \ldots)$, $(d_1, d_2, \ldots)$ are real numbers. $S(\tau)$ is a group that is generated by the operator $(A + B)$. The product of the exponential functions is called the *integrator*,

Thus for the construction of a first-order method, we have the trivial solution:

$$c_1 = d_1 = 1 \text{ and } m = 1. \tag{3.15}$$

For a fourth-order method, see [160], we have the following coefficients:

$$c_1 = c_4 = \frac{1}{2(2 - 2^{1/3})}, c_2 = c_3 = \frac{1 - 2^{1/3}}{2(2 - 2^{1/3})}, \tag{3.16}$$

$$d_1 = d_3 = \frac{1}{2 - 2^{1/3}}, d_2 = -\frac{2^{1/3}}{2 - 2^{1/3}}, d_4 = 0. \tag{3.17}$$

We can improve this direct method by using symmetric integrators and obtain orders of 4, 6, 8, ..., see [199]. For this construction the exact reversibility in time is important, i.e. $S(\tau)S(-\tau) = S(-\tau)S(\tau)$.

REMARK 3.1 The construction of the higher-order methods is based on the forward and backward time steps, due to time reversibility. *Negative time steps are also possible.* A visualization of the splitting steps is presented in Figure 3.2. ▯

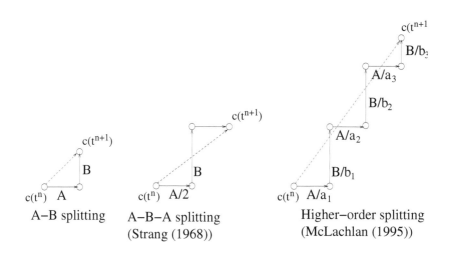

A–B splitting   A–B–A splitting
(Strang (1968))

Higher–order splitting
(McLachlan (1995))

FIGURE 3.2:   Graphical visualization of noniterative splitting methods.

REMARK 3.2 The construction of the higher-order method can also be achieved with respect to the symplectic operators, e.g. conservation of the symplectic two-form $dp \wedge dq$, see [199]. Based on this symplectic behavior, the total energy is guaranteed. ▯

REMARK 3.3 Physical restrictions, as temporal irreversibility, e.g. quantum statistical trace or the imaginary temporal Schröder equation, will also require positive coefficients of higher-order methods. Based on these restrictions, theorems for the higher-order methods have been developed, see [42]. The general idea is to force some coefficients, and thus their underlying commutators, to be zero. ▯

REMARK 3.4 The construction of higher-order methods based on symplectic operators can be accomplished by geometric integrators, see [30], [31], [118] and [155]. Therefore transformations to adequate Hamiltonian systems can be delicate and can require problems. For our abstract parabolic and hyperbolic equation systems, we present an alternate idea for constructing higher-order methods, based on iterative methods, see [69] and [70]. To obtain higher-order methods, each starting solution for the next iterative process and the initial values have to be as accurate as possible, see [69]. The construction can be achieved by sufficient approximate solutions, see the following sections. ▯

3.2 Iterative Operator-Splitting Methods for Bounded Operators

In this chapter, we introduce the modern operator-splitting methods based on the iterative methods. The problem with classical operator-splitting methods is that they decouple equations and moreover physical problems, causing separation of coupled effects. In recent years, new methods have been established for engineering applications, in which equations are decoupled and iterative methods are used to skip the decoupling error, cf. [130] and [134].

3.2.1 Physical Operator-Splitting Methods

The main advantage of decoupling operators in the equations is the computational efficiency in their different temporal and spatial scales with respect to the most accurate discretization and solver methods.

A classical example is to decouple a multiphysics problem (e.g., a diffusion-reaction equation into diffusion and reaction parts), see Figure 3.3. Each part can be solved with its accurate method (e.g., implicit time discretization and finite element methods for the diffusion part and higher-order explicit Runge-Kutta methods for the reaction part).

Decoupled operators can have, for example, stiff and nonstiff, monotone and nonmonotone, or linear and nonlinear behavior. The physical behavior for the decoupling is an important consideration. The physical scales can be an advantage in decomposing the different behaviors of the terms. We could embed the efficiency of controlling the different material behaviors while decoupling in different scales and domains. The scales can be coupled either by iterative

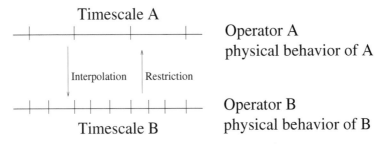

FIGURE 3.3: Decoupling of two physical effects (e.g., a diffusion-reaction equation, where operator A presents the diffusion part and operator B presents the reaction part).

methods or by direct analytical methods.

We can divide the methods into the following two classes of decomposition methods in time:

- Relaxation methods: iterative methods (e.g., iterative operator-splitting methods or waveform-relaxation methods, see [194]).

- Direct (analytical methods): classical operator-splitting methods, the fractional stepping Runge-Kutta (FS-RK), or the stiff backward differential formula (SBDF) method, see [172].

Semi-discretization of a partial differential equation (PDE) leads to an ordinary differential equation (ODE), and therefore we can use the time discretization methods with respect to their different timescales.

By treating both the temporal and spatial scales equivalently, balancing the discretization order of time and space is also possible. By neglecting this consideration, we cannot reach a higher order in time without having a higher order in the space dimension.

3.2.1.1 Example for Decoupling a Differential Equation

We concentrate on the following model problem, given as a diffusion-reaction equation:

$$\frac{\partial c}{\partial t} = D\frac{\partial^2 c}{\partial x^2} - \lambda\, c, \text{ for } x \in \Omega,\ t \in [0, T], \tag{3.18}$$

$$c(x, 0) = c_0(x), \text{ for } x \in \Omega, \tag{3.19}$$

$$c(x, t) = g(x, t), \text{ for } x \in \partial\Omega,\ t \in [0, T], \tag{3.20}$$

where $D, \lambda \in \mathbb{R}^+$ are constant parameters. $c_0(x)$ is the initial condition, and $g(x, t)$ is a function for the Dirichlet boundary condition.

For the discretization, we use second order in space and higher order in time. With the spatial discretization we get the following equation:

$$\frac{\partial c_i}{\partial t} = \frac{D}{\Delta x^2}(c_{i+1} - 2c_i + c_{i-1}) - \lambda c_i, \tag{3.21}$$

where i is the spatial discretization index of the grid nodes.

We have the following scales: for the diffusion operator, we have $\frac{2D}{\Delta x^2}$, and for the reaction operator we have λ. We decouple both operators, if the conditions $\frac{2D}{\Delta x^2} \ll \lambda$ or $\frac{2D}{\Delta x^2} \gg \lambda$ are met. If the scales are each approximate, we neglect the decoupling.

REMARK 3.5 The scales of the operators can be changed, if we assume a higher-order discretization in space or finer spatial grids. Therefore, the balance between the order of the time and space discretization is important and should be made efficient. ⬚

3.2.2 Introduction to the Iterative Operator-Splitting Methods

The iterative operator-splitting methods underlie iterative methods used to solve coupled operators by using a fixed-point iteration. These algorithms integrate each underlying equation with respect to the last iterated solution. Therefore, the starting solution in each iterative equation is important to guarantee fast convergence or a higher-order method. The last iterative solution should have at least the local error of $\mathcal{O}(\tau^i)$, where i is the number of the iteration steps, to obtain the next higher order.

We deal with at least two equations, and therefore two operators, but the results can be generalized to n operators (see, for example, ideas of the waveform-relaxation methods [194]).

In our next analysis, we deal with the following problem:

$$\frac{dc(t)}{dt} = Ac(t) + Bc(t), \text{ for } 0 \leq t \leq T, \tag{3.22}$$
$$c(0) = c_0,$$

where A, B are bounded linear operators. For such a problem, we can derive the analytical solution, given as

$$c(t) = \exp((A + B)t) \, c_0, \text{ for } 0 \leq t \leq T. \tag{3.23}$$

We propose the iterative operator-splitting method as a decomposition method as an effective solver for large systems of partial differential equations.

The iterative operator-splitting methods belong to a second type of iterative method for solving coupled equations. We can combine the traditional operator-splitting method (decoupling the time interval into smaller parts with the splitting time-step) and the iterative splitting method (on each split time interval we use the one-step iterative methods). At the least, the iterative splitting methods serve as predictor-corrector methods, so in the first equation the solution is predicted, whereas in the second equation the solution is corrected, see [130].

We use the iterative operator-splitting methods, because the traditional operator-splitting has, in addition to its benefits, several drawbacks:

- For noncommuting operators, there may be a very large constant in the local splitting error, requiring the use of an unrealistically small splitting time-step. In other words, the stability and commutativity are connected by the norm of the commutator, see Remark 3.6.

- Within a full splitting step in one subinterval, the inner values are not an approximation to the solution of the original problem.

- Splitting the original problem into the different subproblems with one operator (i.e., neglecting the other components) is physically correct, see for example the Strang splitting. But the method is physically questionable, when we aim to get consistent approximations after each inner step, because we lose the exact starting conditions.

Thus, for the iterative splitting methods, we state the following theses:

- For noncommuting operators, we may reduce the local splitting error by using more iteration steps to obtain a higher-order accuracy.

- We must solve the original problem within a full splitting step, while keeping all operators in the equations.

- Splitting the original problem into the different subproblems, including all operators of the problem, is physically the best. We obtain consistent approximations after each inner step because of the exact or approximate starting conditions for the previous iterative solution.

REMARK 3.6 The commutator is related to the consistency of the method. We assume the sequential splitting method:

$$\frac{dc_1(t)}{dt} = Ac_1(t), \ c_1(t^n) = c^n, \tag{3.24}$$

$$\frac{dc_2(t)}{dt} = Bc_2(t), \ c_2(t^n) = c_1(t^{n+1}), \tag{3.25}$$

where A, B are bounded operators in a Banach space $\mathbf{X}$. The time-step is $\tau = t^{n+1} - t^n$, with $t \in [t^n, t^{n+1}]$. The result for the splitting method is given as $c(t^{n+1}) \approx c_2(t^{n+1}) = c_{sp}(t^{n+1})$.

Then we obtain the local error:

$$||\text{err}_{\text{local}}(\tau)|| = ||(\exp((A+B)\tau) - \exp(A\tau)\exp(B\tau))c_0||, \tag{3.26}$$
$$\leq ||[A,B]||\mathcal{O}(\tau^2), \tag{3.27}$$

where $\text{err}_{\text{local}}(\tau) = c(t^{n+1}) - c_{sp}(t^{n+1})$ is defined, and for the stability, the commutator in the norm $|| \cdot ||_{\mathbf{X}} = || \cdot ||$ (e.g., the maximum norm), must be bounded (e.g., $||[A,B]|| < C$, where C is a constant). ☐

In order to avoid the problems mentioned above, we can use the iterative operator-splitting on the interval $[0,T]$, cf. [130]. In the following discussion, we suggest a modification of this method by introducing the splitting time discretization. We suggest an algorithm, based on the iteration for the fixed sequential operator-splitting discretization with the step size τ_n. On the time interval $[t^n, t^{n+1}]$, we solve the following subproblems consecutively for $i = 1, 3, 5, \ldots 2m + 1$:

$$\frac{dc_i(t)}{dt} = Ac_i(t) + Bc_{i-1}(t), \ \text{with} \ c_i(t^n) = c^n_{sp}, \tag{3.28}$$

$$\frac{dc_{i+1}(t)}{dt} = Ac_i(t) + Bc_{i+1}(t), \ \text{with} \ c_{i+1}(t^n) = c^n_{sp}, \tag{3.29}$$

where $c_0(t)$ is any fixed function for each iteration (e.g., $c_0(t) = 0$). (Here, as before, c_{sp}^n denotes the known split approximation at the time level $t = t^n$.) The split approximation at the time level $t = t^{n+1}$ is defined as $c_{\text{sp}}^{n+1} = c_{2m+2}(t^{n+1})$.

The algorithm (3.28)-(3.29) is an iterative method, which at each step consists of both operators A and B. Hence, in these equations, there is no real separation of the different physical processes. However, we note that subdividing the time interval into subintervals distinguishes this process from the simple fixed-point iteration and turns it into a more efficient numerical method.

We also observe that the algorithm (3.28)-(3.29) is a real operator-splitting method, because Equation (3.28) requires a problem with the operator A to be solved, and (3.29) requires a problem with the operator B to be solved. Hence, as in the sequential operator-splitting, the two operators are separated.

3.2.3 Consistency Analysis of the Iterative Operator-Splitting Method

In this subsection, we analyze the consistency and the order of the iterative operator-splitting method. First, in Section 3.2.3.1 we consider the original algorithm (3.28)-(3.29), prove its consistency, and define the order of the local splitting error.

The algorithm (3.28)–(3.29) requires the knowledge of the functions $c_{i-1}(t)$ and $c_i(t)$ on the whole interval $[t^n, t^{n+1}]$, which is typically not the case, because generally their values are known only at several points of the split interval. Hence, typically we can define only some interpolations of these functions. In Section 3.2.3.2, we prove the consistency of such a modified algorithm.

3.2.3.1 Local Error Analysis of the Iterative Operator-Splitting Method

Here we will analyze the consistency and the order of the local splitting error of the method (3.28)–(3.29) for the linear bounded operators $A, B : \mathbf{X} \to \mathbf{X}$, where $\mathbf{X}$ is a Banach space. In the following, we use the notation $\mathbf{X}^2$ for the product space $\mathbf{X} \times \mathbf{X}$ supplied with the norm $\|(u, v)^T\| = \max\{\|u\|, \|v\|\}$ $(u, v \in \mathbf{X})$.

We have the following consistency order of our iterative operator-splitting method.

THEOREM 3.1

Let $A, B \in \mathcal{L}(\mathbf{X})$ be given linear bounded operators. We consider the abstract

Cauchy problem:

$$\partial_t c(t) = Ac(t) + Bc(t), \quad 0 < t \leq T,$$
$$c(0) = c_0. \tag{3.30}$$

Then the problem (3.30) has a unique solution. The iteration (3.28)–(3.29) for $i = 1, 3, \ldots, 2m + 1$ is consistent with the order of the consistency $\mathcal{O}(\tau_n^{2m+1})$.

PROOF

Because $A + B \in \mathcal{L}(\mathbf{X})$, it is a generator of a uniformly continuous semigroup; hence, the problem (3.30) has a unique solution $c(t) = \exp((A + B)t)c_0$.

Let us consider the iteration (3.28)–(3.29) on the subinterval $[t^n, t^{n+1}]$. For the local error function $e_i(t) = c(t) - c_i(t)$, we have the following relations:

$$\partial_t e_i(t) = Ae_i(t) + Be_{i-1}(t), \quad t \in (t^n, t^{n+1}],$$
$$e_i(t^n) = 0, \tag{3.31}$$

and

$$\partial_t e_{i+1}(t) = Ae_i(t) + Be_{i+1}(t), \quad t \in (t^n, t^{n+1}],$$
$$e_{i+1}(t^n) = 0, \tag{3.32}$$

for $i = 1, 3, 5, \ldots$, with $e_1(0) = 0$ and $e_0(t) = c(t)$. We use the notation $\mathbf{X}^2$ for the product space $\mathbf{X} \times \mathbf{X}$ supplied with the norm $\|(u, v)\| = \max\{\|u\|, \|v\|\}$ $(u, v \in \mathbf{X})$. The elements $\mathcal{E}_i(t)$, $\mathcal{F}_i(t) \in \mathbf{X}^2$ and the linear operator $\mathcal{A} : \mathbf{X}^2 \to \mathbf{X}^2$ are defined as follows:

$$\mathcal{E}_i(t) = \begin{bmatrix} e_i(t) \\ e_{i+1}(t) \end{bmatrix}, \quad \mathcal{F}_i(t) = \begin{bmatrix} Be_{i-1}(t) \\ 0 \end{bmatrix}, \quad \mathcal{A} = \begin{bmatrix} A & 0 \\ A & B \end{bmatrix}. \tag{3.33}$$

Then, using the notations (3.33), the relations (3.31)–(3.32) can be written in the form

$$\partial_t \mathcal{E}_i(t) = \mathcal{A}\mathcal{E}_i(t) + \mathcal{F}_i(t), \quad t \in (t^n, t^{n+1}],$$
$$\mathcal{E}_i(t^n) = 0. \tag{3.34}$$

Because of our assumptions, $\mathcal{A}$ is a generator of the one-parameter C_0 semigroup $(\exp \mathcal{A}t)_{t \geq 0}$. Hence, by using the variations of constants formula, the solution of the abstract Cauchy problem (3.34) with homogeneous initial conditions can be written as

$$\mathcal{E}_i(t) = \int_{t^n}^t \exp(\mathcal{A}(t - s))\mathcal{F}_i(s)ds, \quad t \in [t^n, t^{n+1}]. \tag{3.35}$$

Then, using the denotation

$$\|\mathcal{E}_i\|_\infty = \sup_{t \in [t^n, t^{n+1}]} \|\mathcal{E}_i(t)\|, \tag{3.36}$$

we have

$$\|\mathcal{E}_i(t)\| \le \|\mathcal{F}_i\|_\infty \int_{t^n}^t \|\exp(\mathcal{A}(t-s))\| ds$$

$$= \|B\|\|e_{i-1}\| \int_{t^n}^t \|\exp(\mathcal{A}(t-s))\| ds, \quad t \in [t^n, t^{n+1}].$$

(3.37)

Because $(\mathcal{A}(t))_{t\ge 0}$ is a semigroup, the *growth estimation*,

$$\|\exp(\mathcal{A}t)\| \le K \exp(\omega t); \quad t \ge 0,$$

(3.38)

holds with some numbers $K \ge 0$ and $\omega \in \mathbb{R}$.

- Assume that $(\mathcal{A}(t))_{t\ge 0}$ is a bounded or exponentially stable semigroup; that is, (3.38) holds for some $\omega \le 0$. Then obviously the estimate

$$\|\exp(\mathcal{A}t)\| \le K, \quad t \ge 0,$$

(3.39)

holds, and hence, according to (3.37), we have the relation

$$\|\mathcal{E}_i\|(t) \le K\|B\|\tau_n\|e_{i-1}\|, \quad t \in [t^n, t^{n+1}].$$

(3.40)

- Assume that $(\exp \mathcal{A}t)_{t\ge 0}$ has an exponential growth with some $\omega > 0$. Using (3.38) we have

$$\int_{t^n}^t \|\exp(\mathcal{A}(t-s))\| ds \le K_\omega(t), \quad t \in [t^n, t^{n+1}],$$

(3.41)

where

$$K_\omega(t) = \frac{K}{\omega} \left(\exp(\omega(t - t^n)) - 1 \right), \quad t \in [t^n, t^{n+1}].$$

(3.42)

Hence,

$$K_\omega(t) \le \frac{K}{\omega} \left(\exp(\omega \tau_n) - 1 \right) = K\tau_n + \mathcal{O}(\tau_n^2).$$

(3.43)

The estimations (3.40) and (3.43) result in

$$\|\mathcal{E}_i\|_\infty = K\|B\|\tau_n\|e_{i-1}\| + \mathcal{O}(\tau_n^2).$$

(3.44)

Taking into account the definition of $\mathcal{E}_i$ and the norm $\|\cdot\|_\infty$ (supremum norm), we obtain

$$\|e_i\| = K\|B\|\tau_n\|e_{i-1}\| + \mathcal{O}(\tau_n^2),$$

(3.45)

and consequently,

$$\|e_{i+1}\| = K\|B\|\|e_i\| \int_{t^n}^t \|\exp(\mathcal{A}(t-s))\| ds,$$

$$= K\|B\|\tau_n(K\|B\|\tau_n\|e_{i-1}\| + \mathcal{O}(\tau_n^2)),$$

$$= K_2\tau_n^2\|e_{i-1}\| + \mathcal{O}(\tau_n^3).$$

(3.46)

We apply the recursive argument that proves our statement. ⬚

REMARK 3.7 When A and B are matrices (i.e., when (3.28)–(3.29) is a system of the ordinary differential equations), we can use the concept of the logarithmic norm for the growth estimation (3.38). Hence, for many important classes of matrices, we can prove the validity of (3.38) with $\omega \leq 0$.
⬚

REMARK 3.8 We note that a huge class of important differential operators generate a contractive semigroup. This means that for such problems, assuming the exact solvability of the split subproblems, the iterative splitting method is convergent to the exact solution in second order. ⬚

REMARK 3.9 We note that the assumption $A \in \mathcal{L}(X)$ can be formulated more weakly as it is enough to assume that the operator A is the generator of a C_0 semigroup. ⬚

REMARK 3.10 When T is a sufficiently small number, then we do not need to partition the interval $[0, T]$ into the subintervals. In this case, the convergence of the iteration (3.28)–(3.29) to the solution of the problem (3.30) follows immediately from Theorem 3.1, and the rate of the convergence is equal to the order of the local splitting error. ⬚

REMARK 3.11 Estimate (3.47) shows that after the final iteration step ($i = 2m + 1$), we have the estimation

$$\|e_{2m+1}\| = K_m \|e_0\| \tau_n^{2m} + \mathcal{O}(\tau_n^{2m+1}). \tag{3.47}$$

This relation shows that the constant in the leading term strongly depends on the choice of the initial guess $c_0(t)$. When the choice is $c_0(t) = 0$ (see [130]), then $\|e_0\| = c(t)$ (where $c(t)$ is the exact solution of the original problem) and hence the error might be significant. ⬚

REMARK 3.12 In realistic applications, the final iteration steps $2m + 1$ and the time-step τ_n are chosen in an optimal relation to one another, such that the time-step τ_n can be chosen maximal and with at least three or five iteration steps. Additionally, a final stop criterion as an error bound (e.g., $|c_i - c_{i-1}| \leq$ err with for example err $= 10^{-4}$), helps to restrict the number of steps. A graphical illustration of the iterative splitting method is given in Figure 3.4. ⬚

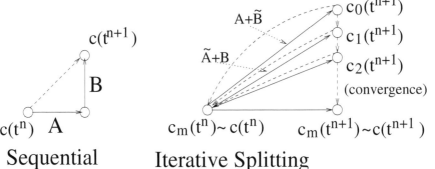

FIGURE 3.4: Graphical visualization of noniterative and iterative splitting methods.

3.2.3.2 Increase of the Order of Accuracy with Improved Initial Functions

We can increase the order of accuracy by improving our choice of the initial iteration function, see [69].

Based on our previous assumption about the initial solutions, we start with exact solutions or an interpolated split solution and present our theory for the exactness of the method.

The Exact Solution of the Split Subproblem
We derive the exact solution of Equations (3.28) and (3.29) by solving the first split problem,

$$c_i(t^{n+1}) = \exp(At)c^n + \sum_{s=0}^{\infty} \sum_{k=s+1}^{\infty} \frac{t^k}{k!} A^{k-s-1} B c_{i-1}^{(s)}(t^n), \qquad (3.48)$$

and the second split problem,

$$c_{i+1}(t^{n+1}) = \exp(Bt)c^n + \sum_{s=0}^{\infty} \sum_{k=s+1}^{\infty} \frac{t^k}{k!} B^{k-s-1} A c_i^{(s)}(t^n), \qquad (3.49)$$

where $\tau = t^{n+1} - t^n$ is the equidistant time-step and $c^n = c(t^n)$ is the exact solution at time t^n or at least approximately of local order $\mathcal{O}(\tau^{m+2})$. n is the number of time-steps ($n \in \{0, \ldots, N\}, N \in \mathbb{N}^+$), and $m > 0$ is the number of iteration steps.

THEOREM 3.2

Assume that for the functions $c_{i-1}(t^{n+1})$ and $c_i(t^{n+1})$ the conditions

$$c_{i-1}^s(t^n) = (A+B)^s c^n, \ s = 0, 1, \ldots, m+1, \tag{3.50}$$
$$c_i^s(t^n) = (A+B)^s c^n, \ s = 0, 1, \ldots, m+2, \tag{3.51}$$

are satisfied. After $m+2$ iterations, the method has a local splitting error $\mathcal{O}(\tau^{m+2})$, and therefore the global error $\mathrm{err_{global}}$ is $\mathcal{O}(\tau^{m+1})$.

PROOF

We show that

$$\exp(\tau(A+B))c^n - c_{m+1}(t^{n+1}) = \mathcal{O}(\tau^{m+1}), \tag{3.52}$$
$$\exp(\tau(A+B))c^n - c_{m+2}(t^{n+1}) = \mathcal{O}(\tau^{m+2}). \tag{3.53}$$

Using the assumption and the exact solutions (3.48) and (3.49), we must proof the relations:

$$\sum_{p=0}^{m+1} \frac{1}{p!}\tau^p (A+B)^p = \sum_{p=0}^{m+1} \frac{1}{p!}\tau^p (A)^p + \sum_{s=0}^{m}\sum_{k=s+1}^{m+1} \frac{\tau^k}{k!} A^{k-s-1} B, \tag{3.54}$$

and

$$\sum_{p=0}^{m+2} \frac{1}{p!}\tau^p (A+B)^p = \sum_{p=0}^{m+2} \frac{1}{p!}\tau^p (B)^p + \sum_{s=0}^{m+1}\sum_{k=s+1}^{m+2} \frac{\tau^k}{k!} B^{k-s-1} A. \tag{3.55}$$

For the proof, we can use the mathematical induction, see [69].

So for each further iteration step, we conserve the order $\mathcal{O}(\tau^{m+1})$ for Equation (3.54) or $\mathcal{O}(\tau^{m+2})$ for Equation (3.55).

We assume for all local errors the order $\mathcal{O}(\tau^{m+2})$.

Based on this assumption, we obtain for the global error

$$\mathrm{err_{global}}(t^{n+1}) = (n+1)\,\mathrm{err_{local}}(\tau)$$
$$= (n+1)\,\tau\,\frac{\mathrm{err_{local}}(\tau)}{\tau} = \mathcal{O}(\tau^{m+1}), \tag{3.56}$$

where we assume equidistant time-steps, a time $t^{n+1} = (n+1)\,\tau$, and the same local error for all $n+1$ time-steps, see also [145].

$\square$

REMARK 3.13 The exact solution of the split subproblem can also be extended to singular perturbed problems and unbounded operators. In these cases, a formal solution with respect to the asymptotic convergence of a power series, which is near the exact solution, can be sought, see [8] and [9].

$\square$

Consistency Analysis of the Iterative Operator-Splitting Method with Interpolated Split Solutions

The algorithm (3.28)–(3.29) requires the knowledge of the functions $c_{i-1}(t)$ and $c_i(t)$ on the whole interval $[t^n, t^{n+1}]$. However, when we solve the split subproblems, usually we apply some numerical methods that allow us to know the values of the above functions only at some points of the interval. Hence, typically we can define only some interpolations to the exact functions.

In the following we consider and analyze the modified iterative process

$$\frac{\partial c_i(t)}{\partial t} = Ac_i(t) + Bc_{i-1}^{\text{int}}(t), \text{ with } c_i(t^n) = c_{\text{sp}}^n, \tag{3.57}$$

$$\frac{\partial c_{i+1}(t)}{\partial t} = Ac_i^{\text{int}}(t) + Bc_{i+1}(t), \text{ with } c_{i+1}(t^n) = c_{\text{sp}}^n, \tag{3.58}$$

where $c_k^{\text{int}}(t)$ (for $k = i-1, i$) denotes an approximation of the function $c_k(t)$ on the interval $[t^n, t^{n+1}]$ with the accuracy $\mathcal{O}(\tau_n^p)$. (For simplicity, we assume the same order of accuracy with the order p on each subinterval.)

The iteration (3.57)–(3.58) for the error function $\mathcal{E}_i(t)$ recalls relation (3.33) with the modified right side; namely,

$$\mathcal{F}_i(t) = \begin{bmatrix} Be_{i-1}(t) + Bh_{i-1}(t) \\ Ah_i(t) \end{bmatrix}, \tag{3.59}$$

where $h_k(t) = c_k(t) - c_k^{\text{int}}(t) = \mathcal{O}(\tau_n^p)$ for $k = i-1, i$. Hence,

$$\|\mathcal{F}_i\|_\infty \le \max\{\|B\| \, \|e_{i-1}\| + \|h_{i-1}\|; \|A\| \, \|h_i\|\}, \tag{3.60}$$

which results in the estimation

$$\|\mathcal{F}_i\|_\infty \le \|B\| \, \|e_{i-1}\| + C \, \tau_n^p. \tag{3.61}$$

Consequently, for these assumptions, the estimation (3.45) turns into the following:

$$\|e_i\| \le K(\|B\| \tau_n \|e_{i-1}\| + C \, \tau_n^{p+1}) + \mathcal{O}(\tau_n^2). \tag{3.62}$$

Therefore, for these assumptions the estimation (3.47) takes the modified form:

$$\|e_{i+1}\| \le K_1 \tau_n^2 \|e_{i-1}\| + KC\tau_n^{p+2} + KC\tau_n^{p+1} + \mathcal{O}(\tau_n^3), \tag{3.63}$$

leading to Theorem 3.3.

THEOREM 3.3

Let $A, B \in \mathcal{L}(X)$ be given linear bounded operators and consider the abstract Cauchy problem (3.30). Then for any interpolation of order $p \ge 1$ the iteration (3.57)–(3.58) for $i = 1, 3, \ldots 2m+1$ is consistent with the order of consistency α where $\alpha = \min\{2m - 1, p\}$.

An outline of the proof can be found in [70].

REMARK 3.14 Theorem 3.3 shows that the number of the iterations should be chosen according to the order of the interpolation formula. For more iterations, we expect a more accurate solution. □

REMARK 3.15 As a result, we can use the piecewise constant approximation of the function $c_k(t)$; namely, $c_k^{int}(t) = c_k(t^n) = \text{const}$, which is known from the split solution. In this instance, it is enough to perform only two iterations in the case of a sufficiently small discretization step size. □

REMARK 3.16 The above analysis was performed for the local error. The global error analysis is as usual and leads to the α-order convergence. □

3.2.4 Quasi-Linear Iterative Operator-Splitting Methods with Bounded Operators

We consider the quasi-linear evolution equation:

$$\frac{dc(t)}{dt} = A(c(t))c(t) + B(c(t))c(t), \text{ for } 0 \le t \le T, \quad (3.64)$$

$$c(0) = c_0, \quad (3.65)$$

where $T > 0$ is sufficiently small and the operators $A(c), B(c) : \mathbf{X} \to \mathbf{X}$ are linear and densely defined in the real Banach space $\mathbf{X}$, see [201].

In the following we modify the linear iterative operator-splitting methods to a quasi-linear operator-splitting method, see [92]. Our aim is to linearize the method by using the old solution for the linearized operators.

The algorithm is based on the iteration with a fixed splitting discretization step size τ. On the time interval $[t^n, t^{n+1}]$, we solve the following subproblems consecutively for $i = 1, 3, \ldots, 2m + 1$:

$$\frac{dc_i(t)}{dt} = A(c_{i-1}(t))c_i(t) + B(c_{i-1}(t))c_{i-1}(t), \text{with } c_i(t^n) = c^n, \quad (3.66)$$

$$\frac{dc_{i+1}(t)}{dt} = A(c_{i-1}(t))c_i(t) + B(c_{i-1}(t))c_{i+1}, \text{with } c_{i+1}(t^n) = c^n, (3.67)$$

where $c_0(t) = 0$ and c^n is the known split approximation at the time level $t = t^n$. The split approximation at the time level $t = t^{n+1}$ is defined as $c^{n+1} = c_{2m+2}(t^{n+1})$. We assume the operators $A(c_{i-1}), B(c_{i-1}) : \mathbf{X} \to \mathbf{X}$ to be linear and densely defined on the real Banach space $\mathbf{X}$, for $i = 1, 3, \ldots, 2m + 1$.

The splitting discretization step size is τ, and the time interval is $[t^n, t^{n+1}]$.

We solve the following subproblems consecutively for $i = 1, 3, \ldots, 2m + 1$:

$$\frac{dc_i(t)}{dt} = \tilde{A}c_i(t) + \tilde{B}(c_{i-1}(t)), \text{ with } c_i(t^n) = c^n, \tag{3.68}$$

$$\frac{dc_{i+1}(t)}{dt} = \tilde{A}(c_i(t)) + \tilde{B}c_{i+1}(t), \text{ with } c_{i+1}(t^n) = c^n, \tag{3.69}$$

where $c_0(t) = 0$ and c^n is the known split approximation at the time level $t = t^n$. The split approximation at the time level $t = t^{n+1}$ is defined as $c^{n+1} = c_{2m+2}(t^{n+1})$. The operators are given as

$$\tilde{A} = A(c_{i-1}), \tilde{B} = B(c_{i-1}). \tag{3.70}$$

We assume bounded operators $\tilde{A}, \tilde{B} : \mathbf{X} \to \mathbf{X}$, where $\mathbf{X}$ is a general Banach space. These operators as well as their sum are generators of the C_0 semigroup. The convergence is examined in a general Banach space setting in the following theorem.

THEOREM 3.4

Let us consider the quasi-linear evolution equation

$$\frac{dc(t)}{dt} = A(c(t))c(t) + B(c(t))c(t), \text{ for } 0 \leq t \leq T,$$
$$c(t^n) = c_n, \tag{3.71}$$

where $A(c)$ and $B(c)$ are linear and densely defined operators in a Banach space, see [201].

We apply the quasi-linear iterative operator-splitting method (3.66)–(3.67) and obtain a second-order convergence rate:

$$\|e_i\| = K \, \tau_n \, \omega_1 \, \|e_{i-1}\| + \mathcal{O}(\tau_n^2), \tag{3.72}$$

where K is a constant. Further, we assume the boundedness of the linear operators with $\max\{\|A(e_{i-1}(t))\|, \|B(e_{i-1}(t))\|\} \leq \omega_1$ for $t \in [0, T]$ and T being sufficiently small.

We can obtain the result with Lipschitz constants, and we prove the argument by using the semigroup theory.

PROOF

Let us consider the iteration (3.66)–(3.67) on the subinterval $[t^n, t^{n+1}]$. For the error function $e_i(t) = c(t) - c_i(t)$, we have the relations

$$\frac{de_i(t)}{dt} = \tilde{A}e_i(t) + \tilde{B}e_{i-1}(t), \quad t \in (t^n, t^{n+1}],$$
$$e_i(t^n) = 0, \tag{3.73}$$

and

$$\frac{de_{i+1}(t)}{dt} = \tilde{A}e_i(t) + \tilde{B}e_{i+1}(t), \quad t \in (t^n, t^{n+1}],$$

$$e_{i+1}(t^n) = 0,$$

(3.74)

for $i = 1, 3, 5, \ldots$, with $e_1(0) = 0$, $e_0(t) = c(t)$, $\tilde{A} = A(e_{i-1})$, and $\tilde{B} = B(e_{i-1})$.

We can rewrite Equations (3.73)–(3.74) into a system of linear first-order differential equations in the following way. The elements $\mathcal{E}_i(t)$, $\mathcal{F}_i(t) \in \mathbf{X}^2$ and the linear operator $\mathcal{A} : \mathbf{X}^2 \to \mathbf{X}^2$ are defined as follows:

$$\mathcal{E}_i(t) = \begin{bmatrix} e_i(t) \\ e_{i+1}(t) \end{bmatrix}, \quad \mathcal{A} = \begin{bmatrix} \tilde{A} & 0 \\ \tilde{A} & \tilde{B} \end{bmatrix},$$

(3.75)

$$\mathcal{F}_i(t) = \begin{bmatrix} \tilde{B}e_{i-1}(t) \\ 0 \end{bmatrix}.$$

(3.76)

Then, using the notations (3.71), the relations (3.75)–(3.76) can be written in the form

$$\partial_t \mathcal{E}_i(t) = \mathcal{A}\mathcal{E}_i(t) + \mathcal{F}_i(t), \quad t \in (t^n, t^{n+1}],$$

$$\mathcal{E}_i(t^n) = 0,$$

(3.77)

because we have assumed that $\tilde{A}$ and $\tilde{B}$ are bounded and linear operators. Furthermore, we have a Lipschitzian domain, and $\mathcal{A}$ is a generator of the one-parameter C_0 semigroup $(\mathcal{A}(t))_{t \geq 0}$. We also assume that the estimation of our term $\mathcal{F}_i(t)$ holds under the growth conditions.

REMARK 3.17 We can estimate the linear operators $A(e_{i-1})$ and $B(e_{i-1})$ by assuming the maximal accretivity and contractivity given as

$$\|A(e_{i-1})y\|_{\mathbf{X}} \leq w_2 \|y\|_{\mathbf{Y}}, \|B(e_{i-1})y\|_{\mathbf{X}} \leq w_3 \|y\|_{\mathbf{Y}},$$

(3.78)

where we have the embedding $\mathbf{Y} \subset \mathbf{X}$, and w_2, w_3 are constants in $\mathbb{R}^+$. ◻

We can estimate the right-hand side $\mathcal{F}_i(t)$ in the following lemma.

LEMMA 3.1
Let us consider the linear and compact operator $\tilde{B}$. We can then estimate $\mathcal{F}_i(t)$ as follows:

$$\|\mathcal{F}_i(t)\| \leq w_3 \|e_{i-1}(t)\|, \quad \text{for } t \in [0, T],$$

(3.79)

where $e_{i-1}(t)$ is the error of the last iteration step $i - 1$ and w_3 an approximation of $\tilde{B}(e_{i-1}(t))$, see Remark 3.17. ◻

PROOF
We have the norm $\|\mathcal{F}_i(t)\| = \max\{\mathcal{F}_{i_1}(t), \mathcal{F}_{i_2}(t)\}$ over the components of the vector.

We estimate each term:

$$\|\mathcal{F}_{i_1}(t)\| \leq \|\tilde{B}(e_{i-1}(t))\|$$
$$\leq \omega_3 \|e_{i-1}(t)\|, \tag{3.80}$$
$$\|\mathcal{F}_{i_2}(t)\| = 0, \tag{3.81}$$

obtaining the estimation

$$\|\mathcal{F}_i(t)\| \leq \omega_3 \|e_{i-1}(t)\|. \tag{3.82}$$

$\square$

Hence, using the variations of constants formula, the solution of the abstract Cauchy problem (3.77) with homogeneous initial condition can be written as

$$\mathcal{E}_i(t) = \int_{t^n}^t \exp(\mathcal{A}(t-s))\mathcal{F}_i(s)ds, \quad t \in [t^n, t^{n+1}]. \tag{3.83}$$

See, for example, reference [63]. Therefore, using the denotation

$$\|\mathcal{E}_i\|_\infty = \sup_{t \in [t^n, t^{n+1}]} \|\mathcal{E}_i(t)\|, \tag{3.84}$$

we have

$$\|\mathcal{E}_i\|(t) \leq \|\mathcal{F}_i\|_\infty \int_{t^n}^t \|\exp(\mathcal{A}(t-s))\| ds$$

$$\tag{3.85}$$

$$= \omega_3 \|e_{i-1}\| \int_{t^n}^t \|\exp(\mathcal{A}(t-s))\| ds, \quad t \in [t^n, t^{n+1}].$$

Because $(\mathcal{A}(t))_{t\geq 0}$ is a semigroup, the *growth estimation*:

$$\|\exp(\mathcal{A}t)\| \leq K \exp(\omega_1 t), \quad t \geq 0, \tag{3.86}$$

holds with some numbers $K \geq 0$ and $\omega_1 = \max\{\omega_2, \omega_3\} \in \mathbb{R}$, see Remark 3.17 and [63].

Because of $\omega_1 \geq 0$, we assume that $(\mathcal{A}(t))_{t\geq 0}$ has an exponential growth width. Using (3.86) we have

$$\int_{t^n}^{t^{n+1}} \|\exp(\mathcal{A}(t-s))\| ds \leq K_{\omega_1}(t), \quad t \in [t^n, t^{n+1}], \tag{3.87}$$

where

$$K_{\omega_1}(t) = \frac{K}{\omega_1} \left(\exp(\omega_1(t-t^n)) - 1 \right), \quad t \in [t^n, t^{n+1}], \tag{3.88}$$

and hence,

$$K_{\omega_1}(t) \leq \frac{K}{\omega_1} \left(\exp(\omega_1 \tau_n) - 1 \right) = K \tau_n + \mathcal{O}(\tau_n^2). \tag{3.89}$$

Thus, the estimations (3.79) and (3.89) result in

$$\|\mathcal{E}_i\|_\infty = K\tau_n\|e_{i-1}\| + \mathcal{O}(\tau_n^2). \tag{3.90}$$

Taking into account the definition of $\mathcal{E}_i$ and the norm $\|\cdot\|_\infty$, we obtain

$$\|e_i\| = K\tau_n\|e_{i-1}\| + \mathcal{O}(\tau_n^2), \tag{3.91}$$

where $K = \omega_1\,\omega_3 \in \mathbb{R}^+$. This proves our statement.

$\square$

3.2.5 Stability Analysis for the Linear and Quasi-Linear Operator-Splitting Method

We concentrate on a stability analysis for the linear ordinary differential equations with commutative operators. Our goal is to derive the general case with the eigenvalues of the operators and apply the recursive computation of the underlying differential equations. In the next section, we discuss the stability of the weighted iterative splitting method, which is an upper class of iterative splitting methods. These upper-class methods can stabilize the iterative operator-splitting methods for stiff problems, see [89].

3.2.5.1 Stability for the Linear Weighted Iterative Splitting Method

The proposed unsymmetrical weighted iterative splitting method is a combination between the sequential splitting method, see [68], and the iterative operator-splitting method, see [70]. The weighting factor w is used as an adaptive switch between lower- and higher-order splitting methods. The following algorithm is based on an iteration with a fixed equidistant step size τ for the splitting discretization. On the time interval $[t^n, t^{n+1}]$, we solve the following subproblems consecutively for $i = 1, 3, \ldots 2m + 1$:

$$\frac{\partial c_i(t)}{\partial t} = Ac_i(t) + w\,Bc_{i-1}(t), \text{ with } c_i(t^n) = c^n, \tag{3.92}$$
$$\text{and } c_i(t^n) = c^n,$$
$$\frac{\partial c_{i+1}(t)}{\partial t} = w\,Ac_i(t) + Bc_{i+1}(t), \tag{3.93}$$
$$\text{with } c_{i+1}(t^n) = w\,c^n + (1-w)\,c_i(t^{n+1}),$$

where $c_0(t) = 0$ and c^n is the known split approximation at the time level $t = t^n$. The split approximation at the time level $t = t^{n+1}$ is defined as $c^{n+1} = c_{2m+2}(t^{n+1})$. For parameter w it holds that $w \in [0,1]$. For $w = 0$, we use the sequential splitting, and for $w = 1$ we use the iterative splitting method, cf. [70].

Because of weighting between the sequential splitting and iterative splitting methods, the initial conditions are also weighted. Thus, we have the final results of the first equation (3.92) appearing in the initial condition for the second equation (3.93).

Damped Iterative Splitting Method

The damped iterative splitting method is the next stable splitting method. In this version, we concentrate on the examples with very stiff operators such as the B-operator. For initial solutions that are far away from the local solution, strong oscillations occur, see [89] and [126]. For this reason, we damp the B-operator in such a way that we relax the initial steps with factors $w \approx 0$.

The following algorithm is based on an iteration with a fixed step size τ for the splitting discretization. On the time interval $[t^n, t^{n+1}]$ we solve the following subproblems consecutively for $i = 1, 3, \ldots 2m + 1$:

$$\frac{\partial c_i(t)}{\partial t} = 2(1 - w)Ac_i(t) + 2w \, Bc_{i-1}(t), \text{ with } c_i(t^n) = c^n, \quad (3.94)$$

and $c_i(t^n) = c^n$,

$$\frac{\partial c_{i+1}(t)}{\partial t} = 2w \, Ac_i(t) + 2(1 - w)Bc_{i+1}(t), \quad (3.95)$$

with $c_{i+1}(t^n) = c^n$,

where $c_0(t) = 0$ and c^n is the known split approximation at the time level $t = t^n$. For our parameter w it holds that $w \in (0, 1/2]$. For $0 < w < 1/2$, we use the damped method and solve only the damped operators. For $w = 1/2$ we use the iterative splitting method, cf. [70].

As discussed earlier for the unsymmetric weighted iterative splitting case, weighting between the sequential splitting and iterative splitting method requires that the initial conditions be also weighted. Consequently, we have the damped version results of the weighted operators for the first equation (3.94) and second equation (3.95), where the initial conditions are equal. This is called the symmetric weighted iterative splitting case and symmetrization of the operators, see [71], [90] and [91].

REMARK 3.18 The theoretical background of the proofs is based on the reformulation with simpler scalar equations, performed with eigenvalue problems. This reformulation allows our stability analysis to be accomplished with the semigroup theory and the A-stability formulation of the ODEs. ⬛

Recursion for the Stability Results

First, we concentrate on the (unsymmetric) weighted iterative splitting method, (3.92) and (3.93). For an overview, we treat the particular case of the initial values with $c_i(t^n) = c^n$ and $c_{i+1}(t^n) = c^n$ for an overview. The general case $c_{i+1}(t^n) = wc^n + (1 - w)c_i(t^{n+1})$ can be treated in the same manner.

We consider the suitable vector norm $|| \cdot ||$ being on $\mathbb{R}^M$, together with its induced operator norm. The matrix exponential of $Z \in \mathbb{R}^{M \times M}$ is denoted by $\exp(Z)$. We assume that there holds:

$$|| \exp(\tau \, A)|| \leq 1 \quad \text{and} \quad || \exp(\tau \, B)|| \leq 1 \quad \text{for all } \tau > 0.$$

It can be shown, that the system (3.92)–(3.93) implies $\| \exp(\tau (A + B)) \| \leq 1$ and is itself stable.

For the linear problem (3.92)–(3.93), it follows by integration that there holds

$$c_i(t) = \exp((t - t^n)A)c^n + \int_{t^n}^{t} \exp((t - s)A) \, \omega \, Bc_{i-1}(s) \, ds, \quad (3.96)$$

and

$$c_{i+1}(t) = \exp((t - t^n)B)c^n + \int_{t^n}^{t} \exp((t - s)B) \, \omega \, Ac_i(s) \, ds. \quad (3.97)$$

With elimination of c_i, we get

$$c_{i+1}(t) = \exp((t - t^n)B)c^n + \omega \int_{t^n}^{t} \exp((t - s)B) \, A \, \exp((s - t^n)A) \, c^n \, ds$$

$$+\omega^2 \int_{s=t^n}^{t} \int_{s'=t^n}^{s} \exp((t - s)B) \, A \, \exp((s - s')A) \, B \, c_{i-1}(s') \, ds' \, ds. \quad (3.98)$$

For the following commuting case, we can evaluate the double integral $\int_{s=t^n}^{t} \int_{s'=t^n}^{s}$ as $\int_{s't^n}^{t} \int_{s=s'}^{t}$ and can derive the weighted stability theory.

Commuting Operators

For more transparency of the formula (3.98), we consider a well-conditioned system of eigenvectors and eigenvalues λ_1 of A and λ_2 of B instead of the operators A and B themselves. Replacing the operators A and B by λ_1 and λ_2, respectively, we obtain after some calculations

$$c_{i+1}(t) = c^n \frac{1}{\lambda_1 - \lambda_2} \left(\omega \lambda_1 \exp((t - t^n)\lambda_1) + ((1 - \omega)\lambda_1 - \lambda_2) \exp((t - t^n)\lambda_2) \right)$$

$$+ c^n \, \omega^2 \frac{\lambda_1 \lambda_2}{\lambda_1 - \lambda_2} \int_{s=t^n}^{t} \left(\exp((t - s)\lambda_1) - \exp((t - s)\lambda_2) \right) ds. \quad (3.99)$$

Note that this relation is commutative in λ_1 and λ_2; hence, the scalar commutator is given as $[\lambda_1, \lambda_2] = [\lambda_2, \lambda_1]$.

Strong Stability

We define $z_k = \tau \lambda_k$, $k = 1, 2$, $\tau = t - t^n$. We start with $c_0(t) = c^n$ and obtain

$$c_{2m}(t^{n+1}) = S_m(z_1, z_2) \, c^n, \quad (3.100)$$

where S_m is the stability function of the scheme with m iterations. We use Equation (3.99) and obtain after some calculations for the first iteration step

$$S_1(z_1, z_2) \, c^n = \omega^2 \, c^n + \frac{\omega \, z_1 + \omega^2 \, z_2}{z_1 - z_2} \exp(z_1) \, c^n \quad (3.101)$$

$$+ \frac{(1 - \omega - \omega^2) \, z_1 - z_2}{z_1 - z_2} \exp(z_2) \, c^n,$$

and for the second iteration step

$$S_2(z_1, z_2)\, c^n = \omega^4\, c^n + \frac{\omega\, z_1 + \omega^4\, z_2}{z_1 - z_2}\, \exp(z_1)\, c^n \tag{3.102}$$

$$+ \frac{(1 - \omega - \omega^4)\, z_1 - z_2}{z_1 - z_2}\, \exp(z_2)\, c^n$$

$$+ \frac{\omega^2\, z_1\, z_2}{(z_1 - z_2)^2}\, ((\omega z_1 + \omega^2 z_2) \exp(z_1)$$

$$+ (-(1 - \omega - \omega^2) z_1 + z_2) \exp(z_2))\, c^n$$

$$+ \frac{\omega^2\, z_1\, z_2}{(z_1 - z_2)^3}\, ((-\omega z_1 - \omega^2 z_2)(\exp(z_1) - \exp(z_2))$$

$$+ ((1 - \omega - \omega^2) z_1 - z_2)(\exp(z_1) - \exp(z_2)))\, c^n.$$

Let us consider the stability given by the following eigenvalues in a wedge:

$$\mathcal{W}_\alpha = \{\zeta \in \mathbb{C} : |\arg(\zeta)| \le \alpha\}. \tag{3.103}$$

For the stability we have $|S_m(z_1, z_2)| \le 1$, whenever $z_1, z_2 \in \mathcal{W}_{\pi/2}$.
The stability of the two iterations is given in the following theorem.

THEOREM 3.5
We have the following stability for the weighted iterative splitting method,
(3.92)–(3.93).
For S_1 we have a strong stability with

$$\max_{z_1 \le 0, z_2 \in \mathcal{W}_\alpha} |S_1(z_1, z_2)| \le 1, \ \forall\, \alpha \in [0, \pi/2] \ with \ 0 \le \omega \le 1.$$

For S_2 we have a strong stability with

$$\max_{z_1 \le 0, z_2 \in \mathcal{W}_\alpha} |S_2(z_1, z_2)| \le 1, \ \forall\, \alpha \in [0, \pi/2] \ with \ 0 \le \omega \le \left(\frac{1}{8\,\tan^2(\alpha) + 1}\right)^{1/8}.$$

PROOF

We consider a fixed $z_1 = z$, $Re(z) < 0$, and $z_2 \to -\infty$. Then we obtain

$$S_1(z, -\infty) = \omega^2(1 - \exp(z)), \tag{3.104}$$

and

$$S_2(z, -\infty) = \omega^4(1 - (1 - z)\exp(z)). \tag{3.105}$$

If $z = x + iy$, $x < 0$, then it holds that

1) For S_1 we have

$$|S_1(z, -\infty)|^2 = \omega^4|(1 - 2\exp(x)\cos(y) + \exp(2x))|, \qquad (3.106)$$

and hence,

$$|S_1(z, -\infty)| \leq 1 \Leftrightarrow \omega^4 \leq \left|\frac{1}{1 - 2\exp(x)\cos(y) + \exp(2x)}\right|. \qquad (3.107)$$

Because of $x < 0$ and $y \in \mathbb{R}$, we can estimate $-2 \leq 2\exp(x)\cos(y)$, and $\exp(2x) \geq 0$.
From (3.107) we obtain $\omega \leq \frac{1}{\sqrt[4]{3}}$.

2) For S_2 we have

$$\begin{aligned}|S_2(z, -\infty)|^2 = \omega^8\{1 - 2\exp(x)[(1 - x)\cos y + y\sin y] \qquad (3.108)\\ + \exp(2x)[(1 - x)^2 + y^2]\},\end{aligned}$$

and after some calculations, we obtain

$$\begin{aligned}|S_2(z, -\infty)| \leq 1 \Leftrightarrow \exp(x) \leq (\frac{1}{\omega^8} - 1)\frac{\exp(-x)}{(1 - x)^2 + y^2}\\ + 2\frac{|1 - x| + |y|}{(1 - x)^2 + y^2}. \qquad (3.109)\end{aligned}$$

We can estimate for $x < 0$, $y \in \mathbb{R}$ $\frac{|1-x|+|y|}{(1-x)^2+y^2} \leq 3/2$, and $\frac{1}{2\tan^2(\alpha)} < \frac{\exp(-x)}{(1-x)^2+y^2}$, where $\tan(\alpha) = y/x$.

Finally, we get the bound $\omega \leq \left(\frac{1}{8\tan^2(\alpha)+1}\right)^{1/8}$. $\qquad$ ▯

REMARK 3.19 The stability for the unsymmetric weighted iterative splitting method is stable for the first and second iteration steps. Recursively we can show further A_α-stability for more than two iteration steps with $\alpha \in [0, \pi/2]$, see [89]. $\qquad$ ▯

The stability for the damped iterative operator-splitting method, see Equations (3.94) and (3.95), is given in the following theorem.

THEOREM 3.6
We have the following stability for the damped iterative splitting method, (3.94)–(3.95):

For S_1 we have a strong stability with

$$\max_{z_1 \leq 0, z_2 \in W_\alpha} |S_1(z_1, z_2)| \leq 1, \forall \alpha \in [0, \pi/2],$$

where the damping factor is given as $\omega \leq 1/2$.

PROOF

We consider a fixed $z_1 = z$, $Re(z) < 0$ and $z_2 \to -\infty$. Then we obtain

$$S_1(z, -\infty) = \frac{1 - \omega}{\omega}(1 - e^z). \tag{3.110}$$

If $z = x + iy$, $x < 0$, then it holds that
For S_1 we have

$$|S_1(z, -\infty)|^2 = \left(\frac{1 - \omega}{\omega}\right)^2 (1 - 2\exp(x)\cos(y) + \exp(2x)), \tag{3.111}$$

and hence,

$$|S_1(z, -\infty)| \leq 1 \Leftrightarrow \left(\frac{1 - \omega}{\omega}\right)^2 \leq \left|\frac{1}{1 - 2\exp(x)\cos y + \exp(2x)}\right|. \tag{3.112}$$

From (3.112) we obtain $0 \leq \omega \leq \frac{1}{2}$.

Recursively we can also show the stability for the next iteration steps.

$\square$

In the next section we derive an *a posteriori* error estimate for the iterative splitting methods, starting with different initial solutions.

3.2.5.2 *A Posteriori* **Error Estimates for the Bounded Operators and the Linear Operator-Splitting Method**

We consider the *a posteriori* error estimates for the beginning time iterations.

We can derive the following theorem for the *a posteriori* error estimates.

THEOREM 3.7

Let us consider the iterative operator-splitting method with bounded operators. c^n is the exact solution at t^n, the time-step is $\tau = t - t^n$. We assume the following different starting solutions for $c_0(t)$ and obtain the a posteriori *error estimates for the first iterations:*
Case 1: $c_0(t) = 0$,

$$c_2 - c_1 = ||B|| \, \tau \, c^n + \mathcal{O}(\tau^2), \tag{3.113}$$

Case 2: $c_0(t) = c^n$,

$$c_2 - c_1 = ||BA + B|| \, \tau^2/2 \, c^n + \mathcal{O}(\tau^3), \tag{3.114}$$

Case 3: $c_0(t) = \exp(B(t-t_n)) \exp(A(t-t_n)) c^n$, *(prestep with A-B splitting method)*

$$c_2 - c_1 = ||[A, B]|| \, \tau^2/2 \, c^n + \mathcal{O}(\tau^3). \tag{3.115}$$

PROOF

We apply the equations (3.96) and (3.97), and show the proof for the first case, $c_0(t) = 0$.

The first iteration c_1 is given as

$$c_1(t) = \exp(A(t - t_n)) \, c^n. \tag{3.116}$$

The second iteration is given as

$$c_2(t) = \exp(B(t - t_n)) \, c^n + \left(\int_{t_n}^{t} \exp(-B(s - t_n))A\exp(A(s - t_n)) \, c^n \, dx \right). \tag{3.117}$$

The Taylor expansion for the two functions leads to

$$c_1(t) = (I + A\tau + \frac{\tau^2}{2!} A^2)c^n + \mathcal{O}(\tau^3), \tag{3.118}$$

and

$$c_2(t) = (I + B\tau + \frac{\tau^2}{2!} B^2 + A\tau + A^2 \frac{\tau^2}{2!} + BA \frac{\tau^2}{2!})c^n + \mathcal{O}(\tau^3). \tag{3.119}$$

Subtracting the approximations we obtain

$$||c_2 - c_1|| \le ||B|| \, \tau \, c^n + \mathcal{O}(\tau^2). \tag{3.120}$$

The same can be done recursively for the next iteration steps.

Based on this proof, we can also derive the *a posteriori* estimates for the other initial cases.

□

REMARK 3.20 The *a posteriori* error estimates are related to the iterative solutions of the method. Therefore, the time-step and the boundedness of the operator B are correlated to the error of the next iteration step. More iteration steps and the nonstiff operator B and therefore a small operator norm, for example, spectral norm, $||B|| \approx \epsilon$ (for example $\epsilon \approx 10^{-6}$) are essential for the reduction of the error.

□

3.2.5.3 Stability for the Quasi-Linear Iterative Splitting Method

The following stability theorem is given for the quasi-linear iterative splitting method, see (3.66)–(3.67).

The convergence is examined in a general Banach space setting and we can prove the following stability theorem.

THEOREM 3.8

Let us consider the quasi-linear differential equation

$$\partial_t c(t) = A(c(t))c(t) + B(c(t))c(t), \text{ for } t^n \leq t \leq t^{n+1},$$
$$c(t^n) = c^n, \tag{3.121}$$

where the operators $A(c), B(c) : \mathbf{X} \to \mathbf{X}$ are linear and densely defined in the real Banach space $\mathbf{X}$, see [201] and Subsection 3.2.4. We can define a norm $|| \cdot || = || \cdot ||_{\mathbf{X}}$ on the Banach space $\mathbf{X}$.
We rewrite the equation (3.121) to obtain

$$\partial_t c(t) = \tilde{A}c(t) + \tilde{B}c(t), \text{ for } t^n \leq t \leq t^{n+1},$$
$$c(t^n) = c^n, \tag{3.122}$$

where $\tilde{A} = A(c)$ and $\tilde{B} = B(c)$ with $\tilde{A}, \tilde{B} : \mathbf{X} \to \mathbf{X}$ are given linear bounded operators being generators of the C_0 semigroup and $c^n \in \mathbf{X}$ is a given element. We assume also

$$||A(c)y||_{\mathbf{X}} \leq \lambda_A ||y||_{\mathbf{Y}}, ||B(c)y||_{\mathbf{X}} \leq \lambda_B ||y||_{\mathbf{Y}}, \tag{3.123}$$

where λ_A, λ_B are constants in $\mathbb{R}^+$ and the embedding $\mathbf{Y} \subset \mathbf{X}$ is assumed.
Then the quasi-linear iterative operator-splitting method (3.66)–(3.67) is stable with the following result:

$$||c_{i+1}(t^{n+1})|| \leq \tilde{K} \, ||c^n|| \, \sum_{j=0}^{i+1} \tau^j \, \lambda_{\max}^j, \tag{3.124}$$

where $\tilde{K} > 0$ is a constant and $c^n = c(t^n)$ is the initial condition, $\tau = (t^{n+1} - t^n)$ is the time-step and $\lambda_{\max}$ is the maximal eigenvalue of the linear and bounded operators $\tilde{A}$ and $\tilde{B}$.

We discuss this theorem in the following proof.

PROOF

Let us consider the iteration (3.66)–(3.67) on the subinterval $[t^n, t^{n+1}]$.
We obtain the eigenvalues of the following linear and bounded operators. Because of the well-posed problem we have, $\lambda_{\tilde{A}}$ eigenvalue of $\tilde{A}$, $\lambda_{\tilde{B}}$ eigenvalue of $\tilde{B}$, see [125] and [201].
Then our iteration methods are given with the eigenvalues as follows:

$$\frac{dc_i(t)}{dt} = \lambda_{\tilde{A}} c_i(t) + \lambda_{\tilde{B}} c_{i-1}(t), \quad t \in (t^n, t^{n+1}],$$
$$c_i(t^n) = c^n, \tag{3.125}$$

and

$$\frac{dc_{i+1}(t)}{dt} = \lambda_{\tilde{A}} c_i(t) + \lambda_{\tilde{B}} c_{i+1}(t), \quad t \in (t^n, t^{n+1}],$$

$$c_{i+1}(t^n) = c^n, \tag{3.126}$$

for $i = 1, 3, 5, \ldots$, with $c^n = c(t^n)$ as the approximated solution of the time level $t = t^n$.

The equations can be estimated as follows:

$$c_i(t^{n+1}) = \exp(\lambda_{\tilde{A}} \tau) c^n + \int_{t^n}^{t^{n+1}} \exp(\lambda_{\tilde{A}}(t - s)) \lambda_{\tilde{B}} c_{i-1}(s) ds, \tag{3.127}$$

which leads to

$$||c_i(t^{n+1})|| \leq K_1 ||c^n|| + \tau K_2 \lambda_{\tilde{B}} ||c_{i-1}(t^{n+1})||. \tag{3.128}$$

Further, the second equation can be estimated as

$$c_{i+1}(t^{n+1}) = \exp(\lambda_{\tilde{B}} \tau) c^n + \int_{t^n}^{t^{n+1}} \exp(\lambda_{\tilde{B}}(t - s)) \lambda_{\tilde{A}} c_i(s) ds, \tag{3.129}$$

which can further be estimated as

$$||c_{i+1}(t^{n+1})|| \leq K_3 ||c^n|| + \tau K_4 \lambda_{\tilde{A}} ||c_i(t^{n+1})||. \tag{3.130}$$

With the recursive argument and the maximum of the eigenvalues we can estimate the equations:

$$||c_{i+1}(t^{n+1})|| \leq ||c^n|| \sum_{j=0}^{i+1} K_j \, \tau^j \, \lambda_{\max}^j, \tag{3.131}$$

$$||c_{i+1}(t^{n+1})|| \leq ||c^n|| \, \tilde{K} \sum_{j=0}^{i+1} \tau^j \, \lambda_{\max}^j, \tag{3.132}$$

where $\tilde{K}$ is the maximum of all constants and $\lambda_{\max} = \max\{\lambda_{\tilde{A}}, \lambda_{\tilde{B}}\}$. ▯

REMARK 3.21 We have stability for sufficient small time-steps τ. In addition, accurate estimates can be derived by using the techniques of the mild or weak solutions, see [201]. ▯

3.2.6 Detection of the Correct Splitting Operators

The goal of this section is to detect the correct combination for the operators in the splitting methods.

We propose the idea of using the underlying eigenvalues of the used operators.

For our given linear operator equation,

$$\frac{dc(t)}{dt} = Ac(t) + Bc(t), \text{ for } t \in [0, T], \tag{3.133}$$

$$c(0) = c_0,$$

where $rank(A) = rank(B) = r$, we solve the following eigenvalue problems by assuming that A and B are supposed to be diagonalizable with a well-conditioned eigensystem:

$$A = X^{-1}D_1X, \ D_1 = diag_{k=1}^r(d_{1,k}),$$

$$\max_{k=1}^r(Re(d_{1,k})) \le C_1, ||X|| \le C_1, \ ||X^{-1}|| \le C_1, \tag{3.134}$$

$$B = Y^{-1}D_2Y, \ D_2 = diag_{k=1}^r(d_{2,k}),$$

$$\max_{k=1}^r(Re(d_{2,k})) \le C_2, ||Y|| \le C_2, \ ||Y^{-1}|| \le C_2, \tag{3.135}$$

and C_1 and C_2 are appropriate constants of moderate size. $Re(d_{1,k})$ is the real value of $d_{1,k}$ and $diag_{k=1}^r(d_{1,k})$ is a diagonal matrix with entries $d_{1,k}, \ k \in \{1, \ldots, r\}$.

For the characterization of the different scales of the operators, we apply the maximum of the eigenvalues at each time-step and decide the values for the next time-steps.

We split the operators, if we have

$$\max_{k=1}^r(Re(d_{1,k})) << \max_{k=1}^r(Re(d_{2,k})), \tag{3.136}$$

or

$$\max_{k=1}^r(Re(d_{2,k})) << \max_{k=1}^r(Re(d_{1,k})), \tag{3.137}$$

and our local time-steps for operator A are proposed as $\tau_A < \frac{1}{\max_{k=1}^r(Re(d_{1,k}))}$, and for operator B, the time-steps are proposed as $\tau_B < \frac{1}{\max_{k=1}^r(Re(d_{2,k}))}$.

We do not split if we have

$$\max_{k=1}^r(Re(d_{1,k})) \approx \max_{k=1}^r(Re(d_{2,k})), \tag{3.138}$$

and our time-steps are given for the operators as $\tau_A \approx \tau_B < \frac{1}{\max_{k=1}^r(Re(d_{1,k}))}$.

3.3 Iterative Operator-Splitting Methods for Unbounded Operators

In this section, our goal is to generalize the results of the iterative splitting methods for unbounded operators.

We present our partial differential equations with boundary and initial value conditions as a Cauchy problem of the form

$$\frac{dc}{dt}(t) + Ac(t) = f(t), \ t \in (0, T), \tag{3.139}$$

$$c(0) = c_0, \tag{3.140}$$

where we have an appropriate Hilbert space H, A is a linear unbounded operator in H, and $c_0, f \in L^1(0, T; H)$ are given. The boundary conditions are implicitly incorporated into the domain $D(A)$ of A. Formally, we discuss an ordinary differential equation on the infinite dimensional space H. We run into the problems associated with deriving our results in such infinite dimensional spaces, see [26] and [142].

However, in most applications to partial differential equations, and in our particular case of the convection-diffusion-reaction equations, we can also develop for underlying unbounded operators a complete existence and uniqueness theory for the Cauchy problem, see [13].

Due to favorable properties, our consistency theory of the underlying splitting methods can use such results.

We will briefly describe the basic results of this theory and use these results for a consistency analysis of real-life problems.

The notations of linear accretive operators in Hilbert spaces are essential, and based on this we can derive the estimates of the Cauchy problem.

We have H as a real Hilbert space with the scalar product $(\cdot, \cdot)$ and the norm $|\cdot|$. Let $A : H \to H$ be a linear operator with the domain $D(A)$.

The operator is called *accretive* of $(Ac, c) \geq 0$, $\forall c \in D(A)$, and *m-accretive* if it is accretive and $R(I + A) = H$ (where $R(I + A)$ is the range of $I + A$).

PROPOSITION 3.1

The linear operator A is m-accretive, if and only if for all $\lambda > 0$, $R(I + \lambda A) = H$, $(I + \lambda A)^{-1} \in \mathcal{L}(H)$ and we obtain the error estimates

$$||(I + \lambda A)^{-1})||_{\mathcal{L}(H)} \leq 1, \ \forall \lambda > 0. \tag{3.141}$$

The proof is given in [13].

With this proposition, we can derive the error estimates in the following theorem.

THEOREM 3.9

Let A be an accretive operator and $c_0 \in D(A)$, $f \in C^1([0, T]; H)$ be given. Then the problem (3.139) has a unique solution $c \in C^1([0, T]; H) \cap C([0, T], D(A))$.

The following estimates hold:

$$|c(t)| \leq |c_0| + \int_0^t |f(s)| \, ds, \forall t \in [0, T], \tag{3.142}$$

$$\left| \frac{dc}{dt}(t) \right| \leq |f(0) - Ac_0| + \int_0^t \left| \frac{df(s)}{ds} \right| \, ds, \ t \in (0, T). \tag{3.143}$$

The proof is given in [13].

3.3.1 Consistency Analysis for Unbounded Operators

The consistency analysis is based on the fundamental solutions of the operator equations. In the next subsections, we discuss the examples for different real-life applications in which unbounded operators have such properties, permitting us to develop a consistency theory for the equations.

3.3.1.1 Diffusion-Reaction Equation

However, in most applications to partial differential equations and in particular in the case of the heat equation, the operator A, though unbounded, has some favorable properties that permit us to develop a complete existence theory. We could use the underlying ideas and develop the consistency theory for our iterative splitting methods, see also [26].

At the least, we must estimate the underlying unbounded operators.

The abstract formulation for the heat equation is given as follows:

$$\frac{\partial c}{\partial t} = D\Delta c(x, t) - \lambda c(x, t), \ \forall x \in \mathbb{R}^n, t \in [0, T], \tag{3.144}$$

$$c(x, 0) = c_0(x), \ \forall x \in \mathbb{R}^n, \tag{3.145}$$

where $x = (x_1, \ldots, x_n)^T$ and the unbounded operator is given as $A = D\Delta - \lambda$, where $D \in \mathbb{R}^+$ and $\lambda \in \mathbb{R}^+$ are constants. Further the operator is positive definite, $(Ac, c) \geq 0$ and symmetric, see [115]. The boundary conditions are incorporated into the domain $D(A) = L^2(\mathbb{R}^n)$. The uniqueness and existence are given, see [13], and therefore, if $c_0 \in L^2(\mathbb{R}^n)$, then $c \in C^1((0, T]; L^2(\mathbb{R}^n)) \cap C((0, T]; H^2(\mathbb{R}^n))$.

The solution of Equation (3.144) is given as

$$c(x, t) = \begin{cases} \frac{1}{(2\sqrt{\pi D t})^n} \exp(-\frac{||x||^2}{4Dt}) \exp(-\lambda t), & \text{if } t > 0, x \in \mathbb{R}^n, \\ 0, & \text{if } t \leq 0. \end{cases} \tag{3.146}$$

The solution (3.146) is also called the *fundamental solution*.

We assume splitting of Equation (3.144) into the two operators $A_1 = D\Delta$

and $A_2 = -\lambda$. Our iterative splitting method is given as

$$\frac{\partial c_i}{\partial t} = D\Delta c_i(x,t) - \lambda c_{i-1}(x,t), \ \forall x \in \mathbb{R}^n, t \in [0,T], \qquad (3.147)$$
$$c_i(x,0) = c_0(x), \ \forall x \in \mathbb{R}^n,$$
$$\frac{\partial c_{i+1}}{\partial t} = D\Delta c_i(x,t) - \lambda c_{i+1}(x,t), \ \forall x \in \mathbb{R}^n, t \in [0,T], \qquad (3.148)$$
$$c_{i+1}(x,0) = c_0(x), \ \forall x \in \mathbb{R}^n.$$

Then our semigroups are given as

$$(S_A(t)c_0)(x) = \int_{\mathbb{R}^n} E_A(x - \xi, t)c_0(\xi)d\xi, \qquad (3.149)$$
$$t \geq 0, c_0 \in L^2(\mathbb{R}^n), x \in \mathbb{R}^n,$$

where E_A is the fundamental solution of (3.144), see [13].
For the splitting method, the semigroups are given as

$$(S_{A_1}(t)c_0)(x) = \int_{\mathbb{R}^n} E_{A_1}(x - \xi, t)c_0(\xi)d\xi$$
$$+ \int_0^t \int_{\mathbb{R}^n} E_{A_1}(x - \xi, t - s)A_2 c_{i-1}(\xi, s) \, d\xi ds, \qquad (3.150)$$
$$\forall (x,t) \in \mathbb{R}^n \times [0,T],$$

where E_{A_1} is the fundamental solution of (3.147), and

$$(S_{A_2}(t)c_0)(x) = \int_{\mathbb{R}^n} E_{A_2}(x - \xi, t)c_0(\xi)d\xi$$
$$+ \int_0^t \int_{\mathbb{R}^n} E_{A_2}(x - \xi, t - s)A_1 c_{i-1}(\xi, s) \, d\xi ds, \qquad (3.151)$$
$$\forall (x,t) \in \mathbb{R}^n \times [0,T],$$

where E_{A_2} is the fundamental solution of (3.148).
The fundamental solutions are analytically described in [13].
The following consistency result is based on the higher consistency order of our iterative splitting method, see Theorem 3.3.

THEOREM 3.10
Let us consider heat equation (3.144) with the unbounded linear operator
$A_1 = D\Delta$, the linear operator $A_2 = -\lambda$, and $A = A_1 + A_2$.
The iterative operator-splitting method (3.147)–(3.148) for
$i = 1, 3, \ldots, 2m + 1$ is consistent with the order of the consistency $\mathcal{O}(\tau_n^{2m+1})$
with the following local estimate:

$$\|c(x, t^{n+1}) - c_i(x, t^{n+1})\| \leq \|c_0\|_{L^2(\mathbb{R}^n)} \, \mathcal{O}(\tau_n^i), \qquad (3.152)$$

where $\tau_n = t^{n+1} - t^n$, and c_0 is the initial condition at t^n.

PROOF

Our iterative equations can be solved, and with respect to the semigroup theory we have the following notation:

$$c_i(x, t^{n+1}) = (S_{A_1}^{i-1}(\tau_n)c_0)(x)$$

$$+ \int_{t^n}^{t^{n+1}} (S_{A_1}^{i-1}(t^{n+1} - s)A_2c_{i-1})(x, s) \, ds, \qquad (3.153)$$

where $c_{i-1}(x, t^{n+1}) = (S_A^{i-1}(\tau_n)c_0)(x)$ is given with the consistency order $\|(S_A^{i-1}(\tau_n)c_0)(x) - (S_A(\tau_n)c_0)(x)\|_{L^2(\mathbb{R}^n)} \leq \mathcal{O}(\tau_n^{i-1})\|c_0\|_{L^2(\mathbb{R}^n)}$. We apply the summability, see Theorem 3.2, and obtain

$$c_i(x, t^{n+1}) = (S_A^i(\tau_n)c_0)(x), \qquad (3.154)$$

with the consistency order
$\|(S_A^i(\tau_n)c_0)(x) - (S_A(\tau_n)c_0)(x)\|_{L^2(\mathbb{R}^n)} \leq \mathcal{O}(\tau_n^i)\|c_0\|_{L^2(\mathbb{R}^n)}$.

Further, the same result is given for the next iterative equation:

$$c_{i+1}(x, t^{n+1}) = (S_{A_2}^i(\tau_n)c_0)(x) + \int_{t^n}^{t^{n+1}} S_{A_2}^i(t^{n+1} - s)A_1c_i(x, s) \, ds \qquad (3.155)$$

where $\|c(x, t^{n+1}) - c_i(x, t^{n+1})\|$ has the consistency order $\mathcal{O}(\tau_n^i)$.
We apply also the summability, see Theorem 3.2, and obtain

$$c_{i+1}(x, t^{n+1}) = (S_A^{i+1}(\tau_n)c_0)(x), \qquad (3.156)$$

which has one more order in accuracy.
Thus the consistency of the $i + 1$st iteration step is given as

$$\|c(x, t^{n+1}) - c_{i+1}(x, t^{n+1})\|_{L^2(\mathbb{R}^n)} \leq \mathcal{O}(\tau_n^{i+1})\|c_0\|_{L^2(\mathbb{R}^n)}, \qquad (3.157)$$

where $\|c_0\|_{L^2(\mathbb{R}^n)}$ is the L^2-norm of the initial condition.

$\Box$

REMARK 3.22 As a result, also for unbounded operators with favorable characteristics, for example the fundamental solutions, this relation can be estimated. Based on these fundamental solutions, we can derive the consistency of the iterative operator-splitting method. $\Box$

3.3.1.2 Convection-Diffusion-Reaction Equation

The next application we presented is the convection-diffusion-reaction equation, which also has favorable properties. We can derive fundamental solutions for some special cases, which take on a similar form to the heat equations discussed in the previous section.

A special formulation of the convection-diffusion-reaction equation is given as follows:

$$\frac{\partial c}{\partial t} = -v\frac{\partial c}{\partial x_1} + D\Delta u(x,t) - \lambda c(x,t), \quad \forall x \in \mathbb{R}^n, t \in [0,T], \quad (3.158)$$

$$c(x,0) = c_0(x), \quad \forall x \in \mathbb{R}^n, \quad (3.159)$$

where $x = (x_1,\ldots,x_n)^T$ and the unbounded operator is given as $A = -v\frac{\partial}{\partial x_1} + D\Delta - \lambda$, where $v, D, \lambda \in \mathbb{R}^+$ are constants. As we saw for the heat equation, the operator is positive definite, $(Au,u) \geq 0$, and symmetric, see [115]. The boundary conditions are incorporated into the domain $D(A) = L^2(\mathbb{R}^n)$. The uniqueness and existence are given, see [13], and therefore if $c_0 \in L^2(\mathbb{R}^n)$, then $u \in C^1((0,T]; L^2(\mathbb{R}^n)) \cap C((0,T]; H^2(\mathbb{R}^n))$.

The solution of Equation (3.158) is given as

$$c(x,t) = \begin{cases} \frac{1}{(2\sqrt{\pi Dt})^n} \exp(-\frac{(-2x_1vt+v^2t^2)}{4Dt}). \\ \exp(-\frac{||x||^2}{4Dt}) \exp(-\lambda t), & \text{if } t > 0, x \in \mathbb{R}^n, \quad (3.160) \\ 0, & \text{if } t \leq 0. \end{cases}$$

The solution (3.160) is the *fundamental solution* with respect to the special formulation. A generalization can be made for the convection part, see [21] and [81].

We assume the splitting of the equation (3.158) into the two operators $A_1 = D\Delta$ and $A_2 = -v\frac{\partial}{\partial x_1} - \lambda$. Our iterative splitting method is

$$\frac{\partial c_i}{\partial t} = D\Delta c_i(x,t)$$

$$+(-v\frac{\partial}{\partial x_1} - \lambda)c_{i-i}(x,t), \forall x \in \mathbb{R}^n, t \in [0,T], \quad (3.161)$$

$$c_i(x,0) = c_0(x), \quad \forall x \in \mathbb{R}^n,$$

$$\frac{\partial c_{i+1}}{\partial t} = D\Delta c_i(x,t)$$

$$+(-v\frac{\partial}{\partial x_1} - \lambda)c_{i+i}(x,t), \forall x \in \mathbb{R}^n, t \in [0,T], \quad (3.162)$$

$$c_{i+1}(x,0) = c_0(x), \quad \forall x \in \mathbb{R}^n.$$

Our semigroups are then

$$(S_A(t)c_0) = \int_{\mathbb{R}^n} E_A(x - \xi, t)c_0(\xi)d\xi, \quad (3.163)$$

$$t \geq 0, c_0 \in L^2(\mathbb{R}^n), x \in \mathbb{R}^n,$$

where E_A is the fundamental solution of (3.158), see [13].

The semigroups for the splitting method are

$$(S_{A_1}(t)c_0) = \int_{\mathbb{R}^n} E_{A_1}(x - \xi, t)c_0(\xi)d\xi$$

$$+ \int_0^t \int_{\mathbb{R}^n} E_{A_1}(x - \xi, t - s)A_2 c_{i-1} \, d\xi ds, \ \forall (x,t) \in \mathbb{R}^n \times [0,T], (3.164)$$

where E_{A_1} is the fundamental solution of (3.161), and

$$(S_{A_2}(t)c_0) = \int_{\mathbb{R}^n} E_{A_2}(x - \xi, t)c_0(\xi)d\xi$$

$$+ \int_0^t \int_{\mathbb{R}^n} E_{A_2}(x - \xi, t - s)A_1 c_{i-1} \, d\xi ds, \ \forall (x,t) \in \mathbb{R}^n \times [0,T], (3.165)$$

where E_{A_2} is the fundamental solution of (3.162).

The fundamental solutions for each part are described in [81].

We then have the following consistency result based on the higher consistency order of our iterative splitting method.

THEOREM 3.11

Let us consider the convection-diffusion-reaction equation (3.158) with the unbounded linear operators $A_1 = D\Delta$ and $A_2 = -\frac{\partial}{\partial x_1} - \lambda$.

The iterative operator-splitting method (3.161)–(3.162) for $i = 1, 3, \ldots, 2m + 1$ is consistent with the order of the consistency $\mathcal{O}(\tau_n^{2m+1})$ with the following local estimate:

$$||c(x, t^{n+1}) - c_i(x, t^{n+1})|| \leq ||c_0||_{L^2(\mathbb{R}^n)} \mathcal{O}(\tau_n^i), \tag{3.166}$$

where $\tau_n = t^{n+1} - t^n$, and u_0 is the initial condition at t^n.

PROOF

The iterative equations can be solved, see Theorem 3.2. We have the following notation in semigroups:

$$c_i(x, t^{n+1}) = (S_{A_1}^{i-1}(\tau_n)c_0)(x)$$

$$+ \int_{t^n}^{t^{n+1}} (S_{A_1}^{i-1}(t^{n+1} - s)A_2 c_{i-1})(x, s) \, ds, \tag{3.167}$$

where $c_{i-1}(x, t^{n+1}) = (S_A^{i-1}(\tau_n)c_0)(x)$ with the consistency order $||S_A^{i-1}(\tau_n)c_0(x) - S_A(\tau_n)c_0(x)||_{L^2(\mathbb{R}^n)} \leq \mathcal{O}(\tau_n^{i-1})||c_0||_{L^2(\mathbb{R}^n)}$. See also the summability notation in Theorem 3.2.

Further, the same result is given for the next iterative equation:

$$c_{i+1}(x, t^{n+1}) = (S_{A_2}^i(\tau_n)c_0)(x)$$

$$+ \int_{t^n}^{t^{n+1}} (S_{A_2}^i(t^{n+1} - s)A_1 c_i)(x, s) \, ds, \tag{3.168}$$

where $c_i(x, t^{n+1}) = (S_A^i(t^{n+1})c_0)(x)$ with the consistency order of $\mathcal{O}(\tau_n^i)$.

Thus, the consistency of the $i + 1$st iteration step is given as

$$||c(x, t^{n+1}) - c_{i+1}(x, t^{n+1})||_{L^2(\mathbb{R}^n)} \leq \mathcal{O}(\tau_n^{i+1})||c_0||_{L^2(\mathbb{R}^n)}, \qquad (3.169)$$

where $||c_0||_{L^2(\mathbb{R}^n)}$ is the L^2-norm of the initial condition.

$\square$

REMARK 3.23 The generalization to systems of convection-diffusion-reaction equations can also be accomplished with respect to the derivation of a fundamental solution, see [81]. The consistency results for the iterative operator-splitting method can be applied, see [69]. $\square$

Chapter 4

Decomposition Methods for Hyperbolic Equations

In this chapter, we focus on methods for decoupling multiphysical and multi-dimensional hyperbolic equations. The main concept is to apply the spatial splitting to the second-order derivative in time. The equations can be performed in their second-order derivatives, which saves more resources, as to convert into a system of first-order derivatives, see [13].

The classical splitting methods for hyperbolic equations are the alternating direction (ADI) methods, see [67], [149], and [155].

A larger group of splitting methods for hyperbolic equations with respect to the linear acoustic wave equation is known as the local one-dimensional (LOD) methods. This approach splits the multidimensional operators into local one-dimensional operators and sweeps implicitly over the directions. The methods are discussed in [37], [59], [103], and [133].

In our work, we apply the iterative splitting methods to second-order partial differential equations, see [104]. We discuss the stability and consistency of the resulting methods.

We discuss strategies for decoupling the different operators with respect to the spectrum of the underlying operators, see [193]. We can see similarities to iterative operator-splitting methods for first-order partial differential equations. The theory is given with the sin- and cos-semigroups and the estimates can be done similarly. The functional analysis in this context is performed for the hyperbolic partial differential equations in [11], [13], [199], [201] and some applications are discussed in [125].

The techniques to treat second-order partial differential equations are given in Figure 4.1.

4.1 Introduction for the Splitting Methods

We deal with the second-order Cauchy problem, derived from applying a semi-discretization in space to the Galerkin method introduced in Equation

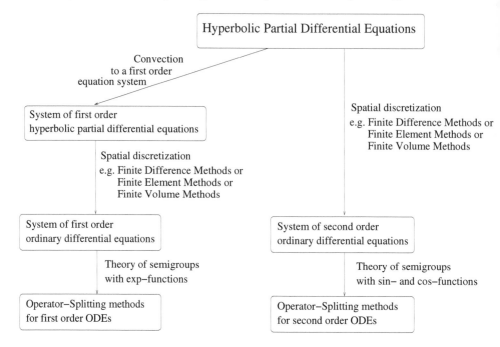

FIGURE 4.1: Applications of the time decomposition methods for hyperbolic equations.

(B.15). We concentrate on the abstract equation

$$\frac{d^2 c(t)}{dt^2} = A_{\text{full}}\, c(t), \text{ for } t \in [0, T]\,, \tag{4.1}$$

where the initial conditions are $c_0 = c(0)$ and $c_1 = \frac{dc}{dt}(0)$. The operator A_{full} is a matrix given with $rank(A_{\text{full}}) = m$, and we assume bounded operators or unbounded operators, see Chapter 3. For example, A_{full} is assembled by the spatial discretization of a Laplace operator, see [13]. Further, we assume the operators to be m-accretive and therefore the unbounded operators can be estimated in some H^d-spaces, see [13]. The solution vector is given as $c(t) = (c_1(t), \ldots, c_m(t))^T$.

We assume a decoupling into two simpler matrices $A_{\text{full}} = A + B$, with $rank(A) = rank(B) = m$.

We also assume that A_{full} can be diagonalized and the two simpler matrices A, B have the same transformation matrix X. We obtain

$$\begin{aligned}
X^{-1} A_{\text{full}} X &= diag_{i=1,\ldots,m}(-\lambda_{A_{\text{full}},i}) \\
&= diag_{i=1,\ldots,m}(-(\lambda_{A,i} + \lambda_{B,i})), \tag{4.2} \\
X^{-1} A X &= diag_{i=1,\ldots,m}(-\lambda_{A,i}), \tag{4.3} \\
X^{-1} B X &= diag_{i=1,\ldots,m}(-\lambda_{B,i}). \tag{4.4}
\end{aligned}$$

Then we can obtain a corresponding system of m independent ordinary differential equations:

$$\frac{d^2 \tilde{c}_i(t)}{dt^2} = -(\lambda_{A,i} + \lambda_{B,i})\, \tilde{c}_i(t), \text{ for } t \in [0, T]\,, \tag{4.5}$$

where $\tilde{c} = X^{-1} c$, $\tilde{c}_{i,0} = \tilde{c}_i(0)$, and $\tilde{c}_{i,1} = \frac{d\tilde{c}_i}{dt}(0)$ are the transformed initial conditions.

We have $\lambda_{A_{\text{full}},i} = \lambda_{A,i} + \lambda_{B,i} \neq 0$ for all $i = 1, \ldots, m$.

We obtain the solution

$$\tilde{c}_i(t) = \tilde{c}_{i,0} \cos(\sqrt{(\lambda_{A,i} + \lambda_{B,i})}t) + \frac{\tilde{c}_{i,1}}{\sqrt{(\lambda_{A,i} + \lambda_{B,i})}} \sin(\sqrt{(\lambda_{A,i} + \lambda_{B,i})}t),$$
$$\text{for } t \in [0, T]\,, \tag{4.6}$$

and all $i = 1, \ldots, m$.

Based on this solution, we can estimate the stability and consistency of our bounded operators in the next section.

For the unbounded and m-accretive operators, we must derive the eigenvalue system and estimate it, for example in the weak formulation, with the H^d-norms, see [13] and [200].

___

4.2 ADI Methods and LOD Methods

In this section, we discuss the traditional splitting methods for hyperbolic equations.
They are based on the finite difference methods in order to derive higher-order methods for the estimation of the Taylor expansion of each discretized operator, with respect to time and space. Classical splitting methods for hyperbolic equations are based on the alternating direction implicit (ADI) methods (see [58], [167]); for the wave equations they are discussed in [67] and [149]. The approach is to formulate difference methods for the wave equation in two or three spatial dimensions and decouple it into simpler equations, for which only tridiagonal systems of linear algebraic equations need to be solved in each time-step. The discretization of each simpler equation is done implicitly or explicitly with respect to the directions.

4.2.1 Classical ADI Method

Our classical method is based on the splitting method of [67] and [149]. We discuss the method for a simple wave equation given as

$$\frac{\partial^2 c(x, y, t)}{\partial t^2} = (A + B)\, c(x, y, t) + f(x, y, t), \text{ for } (x, y, t) \in \Omega \times [0, T], \tag{4.7}$$

$$c(x, y, 0) = c_0(x, y), \frac{\partial c(x, y, 0)}{\partial t} = c_1(x, y), \text{ for } (x, y) \in \Omega, \text{(Initial conditions)},$$

$$c(x, y, t) = 0, \text{ for } (x, y, t) \in \partial\Omega \times (0, T), \text{(Boundary conditions)},$$

where the operators $A = D_1 \frac{\partial^2}{\partial x^2}$ and $B = D_2 \frac{\partial^2}{\partial y^2}$ are given, with $D_1, D_2 \in \mathbb{R}^+$. We assume that c exists and belongs to $C^4(\overline{\Omega} \times [0, T])$.

We denote a grid l for the discretization with the nodes $(i\Delta x, j\Delta y)$, where $\Delta x, \Delta y > 0$ and i, j are integers. Our discretized domain is given as $\overline{\Omega}_h = l \cap \Omega$. For the time discretization we denote $\tau_n = t^{n+1} - t^n$ and $n = 0, 1, \ldots, N$ with $t^{N+1} = T$.

The second-order ADI method is given as, see [149],

$$\tilde{c} - 2c(t^n) + c(t^{n-1}) = \tau_n^2 A(\eta\tilde{c} + (1 - 2\eta)c(t^n) + \eta c(t^{n-1})) \tag{4.8}$$
$$+ \tau_n^2 Bc(t^n)$$
$$+ \tau_n^2 (\eta f(t^{n+1}) + (1 - 2\eta)f(t^n) + \eta f(t^{n-1})),$$
$$\text{in } \Omega_h \times [t^{n-1}, t^{n+1}],$$

$$c(t^{n+1}) - 2c(t^n) + c(t^{n-1}) = \tau_n^2 A(\eta\tilde{c} + (1 - 2\eta)c(t^n) + \eta c(t^{n-1})) \tag{4.9}$$
$$+ \tau_n^2 B(\eta c(t^{n+1}) + (1 - 2\eta)c(t^n) + \eta c(t^{n-1}))$$
$$+ \tau_n^2 (\eta f(t^{n+1}) + (1 - 2\eta)f(t^n) + \eta f(t^{n-1})),$$
$$\text{in } \Omega_h \times [t^{n-1}, t^{n+1}],$$

where the result is given as $c(t^{n+1})$ with the initial conditions $c(t^n) = c^n$, $\frac{dc(t^n)}{dt} = \frac{dc^n}{dt}$ as the known split approximation at the time level $t = t^n$. We have $\eta \in [0, 0.5]$, and the boundary condition is given as $c(x, y, t) = 0$ on the boundary $\partial\Omega_h \times [t^{n-1}, t^{n+1}]$. The local time-step is given as $\tau_n = t^{n+1} - t^n$. The fully coupled explicit method is for $\eta = 0$, and the decoupled implicit method is for $0 < \eta \leq 0.5$, which is a mixing of explicit and implicit Euler methods.

We obtain the following consistency result.

THEOREM 4.1

We apply the second-order discretization in time and space for our underlying ADI method. We then obtain the consistency error

$$|Lc - L_{\tau_n, h}c| = \mathcal{O}(\tau_n^2 + h^2), \tag{4.10}$$

for sufficiently smooth functions $c \in C^4(\overline{\Omega} \times [0,T])$. We have $h = \max\{\Delta x, \Delta y\}$, $Lc = \frac{\partial^2 c}{\partial t^2} - (A+B)c$ is the wave operator (d'Alembert operator), and the discrete operator with respect to the ADI method is given as

$$L_{\tau_n,h}c = \partial_t^+ \partial_t^- c - (D_1 \partial_x^+ \partial_x^- + D_2 \partial_y^+ \partial_y^-)c + R_h(c)$$

$$= \partial_t^+ \partial_t^- c - (D_1 \partial_x^+ \partial_x^- + D_2 \partial_y^+ \partial_y^-)c$$

$$-\eta \tau_n^2 (D_1 \partial_x^+ \partial_x^- + D_2 \partial_y^+ \partial_y^-)\partial_t^+ \partial_t^- c + \eta^2 \tau_n^4 (\partial_t^+ \partial_t^- (D_1 \partial_x^+ \partial_x^- + D_2 \partial_y^+ \partial_y^-)c),$$

where the residual of the ADI method is given as
$$R_h(c) = -\eta \tau_n^2 (D_1 \partial_x^+ \partial_x^- + D_2 \partial_y^+ \partial_y^-)\partial_t^+ \partial_t^- c + \eta^2 \tau_n^4 (\partial_t^+ \partial_t^- (D_1 \partial_x^+ \partial_x^- + D_2 \partial_y^+ \partial_y^-)c),$$
see [149]. Here ∂_p^+ is a forward and ∂_p^- is a backward difference quotient with respect to the variable p.

PROOF

We calculate the consistency error $Lc - L_{\tau_n,h}c$, which is given as

$$|Lc - L_{\tau_n,h}c| = |(\partial_t \partial_t c - (D_1 \partial_x \partial_x + D_2 \partial_y \partial_y)c)$$
$$-(\partial_t^+ \partial_t^- c - (D_1 \partial_x^+ \partial_x^- + D_2 \partial_y^+ \partial_y^-)c + R_h(c))|$$
$$= \mathcal{O}(\tau_n^2 + h^2), \tag{4.11}$$

for $0 < \eta \leq 0.5$. Time and space are discretized as second-order methods, and therefore, we obtain a second-order consistency result, see also [149].

$\square$

REMARK 4.1 Higher-order ADI methods are more complicated to develop, see [67]. Therefore, we concentrate on the LOD method, in which it is possible to derive a fourth-order method. $\square$

4.2.2 LOD Method

The local one-dimensional (LOD) method is discussed in [133].

The intent of this method is to combine explicit and implicit discretization methods to obtain simpler equations and gain a higher-order splitting method, while the splitting error maintains the same order as the discretization method.

The wave equation is given as

$$\frac{\partial^2 c(x,y,t)}{\partial t^2} = (A_1 + A_2)c(x,y,t)$$
$$+ f(x,y,t), \text{ for } (x,y,t) \in \Omega \times (0,T), \tag{4.12}$$
$$c(x,y,0) = c_0(x,y), \frac{\partial c(x,y,0)}{\partial t} = c_1(x,y), \text{ for } (x,y) \in \Omega,$$
$$c(x,y,t) = 0, \text{ for } (x,y,t) \in \partial\Omega \times (0,T),$$

where the unbounded operators are given as $A_1 = D_1 \frac{\partial^2}{\partial x^2}$ and $A_2 = D_2 \frac{\partial^2}{\partial y^2}$, with $D_1, D_2 \in \mathbb{R}^+$.

Further, we assume that the order of the spatial discretization is given by

$$\tilde{A}_1 c \approx A_1 c + \mathcal{O}(h^p), \tag{4.13}$$

$$\tilde{A}_2 c \approx A_2 c + \mathcal{O}(h^p), \tag{4.14}$$

with the order $\mathcal{O}(h^p)$, $p = 2, 4$.

Furthermore, we apply the second-order finite difference method for the discretization in time and we obtain

$$\partial_t^+ \partial_t^- c := \frac{c^{n+1} - 2c^n + c^{n-1}}{\tau_n^2}. \tag{4.15}$$

The LOD method for our Equation (4.12) is then

$$c^{n+1,0} - 2c^n + c^{n-1} = \eta \tau_n^2 (\tilde{A}_1 + \tilde{A}_2)c^n, \tag{4.16}$$

$$c^{n+1,1} - c^{n+1,0} = \eta \, \tau_n^2 \tilde{A}_1 (c^{n+1,1} - 2c^n + c^{n-1}), \tag{4.17}$$

$$c^{n+1} - c^{n+1,1} = \eta \, \tau_n^2 \tilde{A}_2 (c^{n+1} - 2c^n + c^{n-1}), \tag{4.18}$$

for $\eta \in [0, 0.5]$, where the result is given as $c(t^{n+1})$ with the initial conditions $c(t^n) = c^n$, $\frac{dc(t^n)}{dt} = \frac{dc^n}{dt}$ as the known split approximation at the time level $t = t^n$.

Equation (4.16) is the explicit method and gains all operators. The equations (4.17)–(4.18) are implicit and obtain only one operator.

REMARK 4.2 The LOD method obtains the local one-dimensional directions of each operator. Therefore, we can construct unconditional stable methods. ▯

In the next subsection, we discuss the stability and consistency of a higher-order LOD method based on wave equations.

4.2.2.1 Stability and Consistency of the Fourth-Order Splitting Method

The consistency of the fourth-order splitting method is given in the following theorem.

We assume the following discretization orders of $\mathcal{O}(h^p)$, $p = 2, 4$ for the discretization in space, in which $h = h_x = h_y$ is the spatial step size. Further, we assume a second-order time discretization, where τ_n is the equidistant time-step and $t_n = n\tau_n$.

Then we obtain the following consistency of our method (4.16)–(4.18).

THEOREM 4.2
The consistency of our method is given by

$$(\frac{d^2 c(t)}{dt^2} - Ac) - (\partial_t^+ \partial_t^- c - \tilde{A}c) = \mathcal{O}(\tau_n^4), \quad (4.19)$$

where $\partial_t^+ \partial_t^-$ is a second-order discretization in time and $\tilde{A}$ is the discretized fourth-order spatial operator.

PROOF
    We add Equations (4.16)–(4.18) and obtain, see also [133],

$$\partial_t^+ \partial_t^- c^n - \tilde{A}(\theta c^{n+1} + (1 - 2\theta)c^n + \theta c^{n-1})$$
$$-\tilde{B}(c^{n+1} - 2c^n + c^{n-1}) = 0, \quad (4.20)$$

where it holds that

$$\tilde{B} = \theta^2 \tau_n^2 \tilde{A}_1 \tilde{A}_2. \quad (4.21)$$

Therefore, we obtain a splitting error of $\tilde{B}(c^{n+1} - 2c^n + c^{n-1})$.
    For sufficient smoothness we have $(c^{n+1} - 2c^n + c^{n-1}) = \mathcal{O}(\tau_n^2)$, and we obtain $\tilde{B}(c^{n+1} - 2c^n + c^{n-1}) = \mathcal{O}(\tau_n^4)$.
    We thus obtain a fourth-order method, if the spatial operators are also discretized as fourth-order terms.

$\square$

DEFINITION 4.1
    We define the operator $\mathcal{P}^{\pm}(c^j, \theta)$ as

$$\mathcal{P}^{\pm}(c^j, \theta) := \theta(\tilde{A}c^j, c^j) + \theta(\tilde{A}c^{j\pm1}, c^{j\pm1}) + (1 - 2\theta)(\tilde{A}c^j, c^{j\pm1}), \quad (4.22)$$

where $c^j = c(t^j)$ is the solution at time t^j and $\theta \in [0, 0.5]$.

THEOREM 4.3
The stability of the fourth-order splitting method is given by

$$||(1 - \tau_n^2 \tilde{B})^{1/2} \partial_t^+ c^n||^2 + \mathcal{P}^+(c^n, \theta)$$
$$\leq ||(1 - \tau_n^2 \tilde{B})^{1/2} \partial_t^+ c^0||^2 + \mathcal{P}^+(c^0, \theta), \quad (4.23)$$

where $\theta \in [0.25, 0.5]$.

    The proof is given in the following manner.

PROOF
    We must proof the theorem for a test function $\partial_t^0 c^n$, where ∂_t^0 is the central difference.

We choose $\partial_t^0 c^n$ as a test function and for $n \geq 1$ we have

$$
((1 - \tau_n^2 \tilde{B}) \partial_t^+ \partial_t^- c^n, \partial_t^0 c^n)
$$
$$
+ (\tilde{A}(\theta c^{n+1} + (1 - 2\theta)c^n + \theta c^{n-1}), \partial_t^0 c^n) = 0. \tag{4.24}
$$

Multiplying by τ_n and summarizing over j yields

$$
\sum_{j=1}^{n} \left(((1 - \tau_n^2 \tilde{B}) \partial_t^+ \partial_t^- c^j, \partial_t^0 c^j) \, \tau_n \right.
$$
$$
\left. + (\tilde{A}(\theta c^{j+1} + (1 - 2\theta)c^j + \theta c^{j-1}), \partial_t^0 c^j) \, \tau_n \right) = 0. \tag{4.25}
$$

We can derive the identities

$$
((1 - \tau_n^2 \tilde{B}) \partial_t^+ \partial_t^- c^j, \partial_t^0 c^j) \, \tau_n
$$
$$
= 1/2 \|(1 - \tau_n^2 \tilde{B})^{1/2} \partial_t^+ c^j\|^2 - 1/2 \|(1 - \tau_n^2 \tilde{B})^{1/2} \partial_t^- c^j\|^2, \tag{4.26}
$$

$$
(\tilde{A}(\theta c^{j+1} - (1 - 2\theta)c^j + \theta c^{j-1}), \partial_t^0 c^j) \, \tau_n
$$
$$
= 1/2(\mathcal{P}^+(c^j, \theta) - \mathcal{P}^-(c^j, \theta)), \tag{4.27}
$$

and obtain the result

$$
\|(1 - \tau_n^2 \tilde{B})^{1/2} \partial_t^+ c^n\|^2 + \mathcal{P}^+(c^n, \theta)
$$
$$
\leq \|(1 - \tau_n^2 \tilde{B})^{1/2} \partial_t^+ c^0\|^2 + \mathcal{P}^+(c^0, \theta), \tag{4.28}
$$

see also the methodology of [133].

$\square$

REMARK 4.3 For $\theta = \frac{1}{12}$ we obtain a fourth-order method. For the local splitting error, we must use the multiplier $\tilde{A}_1 \tilde{A}_2$; thus, for large constants we have an unconditionally small time-step. $\square$

4.2.3 LOD Method Applied to a System of Wave Equations

Here we discuss the elastic wave propagation given as a system of linear wave equations and introduced in Chapter 2.

4.2.3.1 Basic Numerical Methods

The standard explicit difference scheme with central differences used everywhere was introduced in [4]. To save space, we first give an example in two dimensions.

We discretize uniformly in space and time on the unit square. We then get a grid Ω_h with grid points $x_j = jh, y_k = kh$, where $h > 0$ is the spatial grid size, and a time interval $[0, T]$ with the time points $t_n = n\tau_n$, where τ_n

is the equidistant time-step. The indices are denoted as $j \in \{0, 1, \ldots, J\}$, $k \in \{0, 1, \ldots, K\}$ and $n \in \{0, 1, \ldots, N\}$ with $J, K, N \in \mathbb{N}^+$. If we define the grid function as $U^n_{j,k} = U(x_j, y_k, t_n)$ with $U = (u, v)^T$ or $U = (u, v, w)$, the basic explicit scheme is given as

$$\rho \frac{U^{n+1}_{j,k} - 2U^n_{j,k} + U^{n-1}_{j,k}}{\tau_n^2} = \mathcal{M}_2 U^n_{j,k} + f^n_{j,k}, \qquad (4.29)$$

where $\mathcal{M}_2$ is a difference operator:

$$\mathcal{M}_2 = \begin{pmatrix} (\lambda + 2\mu)\partial_x^2 + \mu\partial_y^2 & (\lambda + \mu)\partial_x^0\partial_y^0 v_{j,k} \\ (\lambda + \mu)\partial_x^0\partial_y^0 v_{j,k} & (\lambda + 2\mu)\partial_y^2 + \mu\partial_x^2 \end{pmatrix}, \qquad (4.30)$$

and we use the standard difference operator notation:

$$\partial_x^+ v_{j,k} = \frac{1}{h}(v_{j+1,k} - v_{j,k}), \quad \partial_x^- v_{j,k} = \partial_x^+ v_{j-1,k}, \quad \partial_x^0 = \frac{1}{2}\left(\partial_x^+ + \partial_x^-\right),$$

and

$$\partial_x^2 = \partial_x^+ \partial_x^-.$$

$\mathcal{M}_2$ approximates $\mu\nabla^2 + (\lambda + \mu)\nabla(\nabla\cdot)$ to second order. This explicit scheme is stable for time-steps satisfying

$$\tau_n < \frac{h}{\lambda + 3\mu}. \qquad (4.31)$$

Replacing $\mathcal{M}_2$ by $\mathcal{M}_4$, we have a fourth-order difference operator given by

$$\mathcal{M}_4 = \begin{pmatrix} (\lambda + 2\mu)\left(1 - \frac{h^2}{12}\partial_x^2\right)\partial_x^2 + \mu\left(1 - \frac{h^2}{12}\partial_y^2\right)\partial_y^2 \\ (\lambda + \mu)\left(1 - \frac{h^2}{6}\partial_x^2\right)\partial_x^0\left(1 - \frac{h^2}{6}\partial_y^2\right)\partial_y^0 \\ (\lambda + \mu)\left(1 - \frac{h^2}{6}\partial_x^2\right)\partial_x^0\left(1 - \frac{h^2}{6}\partial_y^2\right)\partial_y^0 \\ (\lambda + 2\mu)\left(1 - \frac{h^2}{12}\partial_y^2\right)\partial_y^2 + \mu\left(1 - \frac{h^2}{12}\partial_x^2\right)\partial_x^2 \end{pmatrix}, \qquad (4.32)$$

and using the modified equation approach [50], we obtain the explicit fourth-order scheme:

$$\rho \frac{U^{n+1}_{j,k} - 2U^n_{j,k} + U^{n-1}_{j,k}}{\tau_n^2} = \mathcal{M}_4 U^n_{j,k} + f^n_{j,k}$$

$$+ \frac{\tau_n^2}{12}(\mathcal{M}_2 U^n_{j,k} + \mathcal{M}_2 f^n_{i,j} + \partial_{tt} f^n_{i,j}), \qquad (4.33)$$

where $\mathcal{M}_2$ is a second-order approximation to the $\left(\mu\nabla^2 + (\lambda + \mu)\nabla(\nabla\cdot)\right)^2$ operator (i.e., the right-hand-side operator squared). As $\mathcal{M}_2$ only needs to be second-order accurate, it has the same extent in space as $\mathcal{M}_4$, and no more grid points are used. The extra terms on the right-hand side are used to eliminate the second-order error in the approximation of the time derivative, and we obtain a fully fourth-order scheme.

In [133], the following implicit scheme for the scalar wave equation was introduced:

$$\rho \frac{U_{j,k}^{n+1} - 2U_{j,k}^n + U_{j,k}^{n-1}}{\tau_n^2} = \mathcal{M}_4 \left(\theta U_{j,k}^{n+1} + (1 - 2\theta)U_{j,k}^n + \theta U_{j,k}^{n-1} \right) \quad (4.34)$$

$$+ \theta f_{j,k}^{n+1} + (1 - 2\theta) f_{j,k}^n + \theta f_{j,k}^{n-1}.$$

When $\theta = 1/12$, the error of this scheme is fourth order in time and space. For this θ-value, it is, however, only conditionally stable.

In order to make it competitive with the explicit scheme (4.33), we will provide operator-split versions of the implicit scheme (4.34). The presence of the mixed derivative terms that couple different coordinate directions makes creating split versions complicated.

4.2.3.2 Fourth-Order Splitting Method

Here, we present a fourth-order splitting method based on the basic scheme (4.34). We split the operator $\mathcal{M}_4$ into three parts, $\mathcal{M}_{xx}$, $\mathcal{M}_{yy}$, and $\mathcal{M}_{xy}$, where we have

$$\mathcal{M}_{xx} = \begin{pmatrix} (\lambda + 2\mu) \left(1 - \frac{h^2}{12}\partial_x^2\right) \partial_x^2 & 0 \\ 0 & \mu \left(1 - \frac{h^2}{12}\partial_x^+\partial_x^-\right)\partial_x^2 \end{pmatrix},$$

$$\mathcal{M}_{yy} = \begin{pmatrix} \mu \left(1 - \frac{h^2}{12}\partial_y^2\right)\partial_y^2 & 0 \\ 0 & (\lambda + 2\mu)\left(1 - \frac{h^2}{12}\partial_y^2\right)\partial_y^2 \end{pmatrix},$$

$$\mathcal{M}_{xy} = \mathcal{M}_4 - \mathcal{M}_{xx} - \mathcal{M}_{yy}.$$

Our proposed splitting method has the following steps:

$$1. \ \rho \frac{U_{j,k}^* - 2U_{j,k}^n + U_{j,k}^{n-1}}{\tau_n^2} = \mathcal{M}_4 U_{j,k}^n$$

$$+ \theta f_{j,k}^{n+1} + (1 - 2\theta) f_{j,k}^n + \theta f_{j,k}^{n-1}, \quad (4.35)$$

$$2. \ \rho \frac{U_{j,k}^{**} - 2U_{j,k}^*}{\tau_n^2} = \theta \mathcal{M}_{xx} \left(U_{j,k}^{**} - 2U_{j,k}^n + U_{j,k}^{n-1} \right)$$

$$+ \frac{\theta}{2} \mathcal{M}_{xy} \left(U_{j,k}^* - 2U_{j,k}^n + U_{j,k}^{n-1} \right), \quad (4.36)$$

$$3. \ \rho \frac{U_{j,k}^{n+1} - 2U_{j,k}^{**}}{\tau_n^2} = \theta \mathcal{M}_{xx} \left(U_{j,k}^{n+1} - 2U_{j,k}^n + U_{j,k}^{n-1} \right)$$

$$+ \frac{\theta}{2} \mathcal{M}_{xy} \left(U_{j,k}^{**} - 2U_{j,k}^n + U_{j,k}^{n-1} \right). \quad (4.37)$$

Here the first step is explicit, while the second and third steps treat the derivatives along the coordinate axes implicitly and the mixed derivatives explicitly. This is similar to how the mixed case is handled for parabolic problems, see [22].

REMARK 4.4 For the Dirichlet boundary conditions, the splitting method, see (4.35)–(4.37), also conserves the conditions. We can use for the three equations, for U^*, U^{**}, and for U^{n+1}, the same conditions, see (4.35)–(4.37).

For the Neumann boundary conditions and other boundary conditions of higher order, we also have to split the boundary conditions with respect to the splitted operators, see [154]. ⬚

In the next subsection, we discuss the stability and consistency for the LOD method.

4.2.3.3 Stability and Consistency of the Fourth-Order Splitting Method for the System of Wave Equations

The consistency of the fourth-order splitting method is given in the following theorem.

We have for all sufficiently smooth functions $U(x,t)$ the following discretization order:

$$\mathcal{M}_4 U = \mu \nabla^2 U + (\lambda + \mu)\nabla(\nabla \cdot U) + \mathcal{O}(h^4), \tag{4.38}$$

where $\mathcal{O}(h^4)$ is fulfilled for each component.

In the following, we assume the operator $\mathcal{O}$ is vectorial (i.e., the accuracy is given for each component). Furthermore, the split operators are also discretized with the same order of accuracy.

We obtain the following consistency result for the splitting method (4.35)–(4.37).

THEOREM 4.4
The splitting method has a splitting error, which for smooth solutions U is $\mathcal{O}(\tau_n^4)$, where we assume $\tau_n = \mathcal{O}(h)$ as a stability constraint.

PROOF

We assume here that $f = (0,0)^T$. We add the equations (4.35)–(4.37) and obtain, as in the scalar case (see [133]), the following result for the discretized equations:

$$\partial_t^+ \partial_t^- U_{j,k}^n - \mathcal{M}_4(\theta U_{j,k}^{n+1} + (1 - 2\theta)U_{j,k}^n + \theta U_{j,k}^{n-1})$$
$$-\mathcal{N}_{4,\theta}(U_{j,k}^{n+1} - 2U_{j,k}^n + U_{j,k}^{n-1}) = (0,0)^T, \tag{4.39}$$

where $\mathcal{N}_{4,\theta} = \theta^2\tau_n^2(\mathcal{M}_{xx}\mathcal{M}_{yy} + \mathcal{M}_{xx}\mathcal{M}_{xy} + \mathcal{M}_{xy}\mathcal{M}_{yy}) + \theta^3\tau_n^4\mathcal{M}_{xx}\mathcal{M}_{yy}\mathcal{M}_{xy}$. We therefore obtain a splitting error of $\mathcal{N}_{4,\theta}(U_{j,k}^{n+1} - 2U_{j,k}^n + U_{j,k}^{n-1})$.

For sufficient smoothness, we have $(U_{j,k}^{n+1} - 2U_{j,k}^n + U_{j,k}^{n-1}) = \mathcal{O}(\tau_n^2)$, and we obtain $\mathcal{N}_{4,\theta}(U^{n+1} - 2U^n + U^{n-1}) = \mathcal{O}(\tau_n^4)$.

It is important that the influence of the mixed terms can also be discretized with a fourth-order method; therefore, the terms are canceled in the proof.

□

For the stability, we must denote an appropriate norm, which is in our case the $L_2(\Omega)$-norm.

In the following, we introduce the notation of the norms.

REMARK 4.5 For our system, we extend the L_2-norm

$$\|U\|_{L_2}^2 = (U, U)_{L_2} = \int_\Omega U^2 \, dx, \tag{4.40}$$

where $U^2 = u^2 + v^2$ or $U^2 = u^2 + v^2 + w^2$ in two and three dimensions. □

REMARK 4.6 The matrix

$$\mathcal{N}_{4,\theta} = \theta^2\tau_n^2(\mathcal{M}_{xx}\mathcal{M}_{yy} + \mathcal{M}_{xx}\mathcal{M}_{xy} + \mathcal{M}_{xy}\mathcal{M}_{yy})$$
$$+\theta^3\tau_n^4\mathcal{M}_{xx}\mathcal{M}_{yy}\mathcal{M}_{xy}, \tag{4.41}$$

where $\mathcal{M}_{xx}$, $\mathcal{M}_{yy}$ and $\mathcal{M}_{xy}$ are symmetrical and positive-definite matrices. Therefore, the matrix $\mathcal{N}_{4,\theta}$ is also symmetrical and positive-definite. □

Furthermore, we can estimate the norms and define a weighted norm, see [131] and [164].

REMARK 4.7 The energy norm is given as

$$(\mathcal{N}_{4,\theta}U, U)_{L_2} = \int_\Omega (\mathcal{N}_{4,\theta}U \cdot U) \, dx. \tag{4.42}$$

Consequently, we can write

$$\|\mathcal{N}_{4,\theta}U\| \leq \omega\|U\|, \; \forall \, U \in H^d, \tag{4.43}$$

where $\omega \in \mathbb{R}^+$ is the weight and $\mathcal{N}_{4,\theta}$ is bounded. The dimension is d and H is a Sobolev space, so each component of U is in H, see [65]. □

The stability of the fourth-order splitting method is given in the following theorem.

THEOREM 4.5

Let $\theta \in [0.25, 0.5]$, then the implicit time-stepping algorithm, see (4.29), and the split procedure, see (4.35)–(4.37), are unconditionally stable. We can estimate the split procedure iteratively as

$$\|(1 - \tau_n^2 \tilde{\omega})^{1/2} \partial_t^+ U_{j,k}^n\|^2 + \mathcal{P}^+(U_{j,k}^n, \theta) \le \|(1 - \tau_n^2 \tilde{\omega})^{1/2} \partial_t^+ U_{j,k}^0\|^2$$
$$+ \mathcal{P}^+(U_{j,k}^0, \theta), \qquad (4.44)$$

where we have
$$\mathcal{P}^\pm(U_{j,k}^n, \theta) := \theta(\mathcal{M}_4 U_{j,k}^n, U_{j,k}^n) + \theta(\mathcal{M}_4 U_{j,k}^{n\pm1}, U_{j,k}^{n\pm1}) + (1 - 2\theta)(\mathcal{M}_4 U_{j,k}^n, U_{j,k}^{n\pm1}),$$
and
$\mathcal{P}^\pm \ge 0$ *for* $\theta \in [0.25, 0.5]$. *Further,* $1 - \tau_n^2 \tilde{\omega} \in \mathbb{R}^+$ *is the factor for the weighted norm* $(\mathcal{I} - \tau_n^2 \mathcal{N}_{4,\theta})U \le (1 - \tau_n^2 \tilde{\omega}) U$, $\forall U \in H^d$, *where* $\mathcal{I}$ *is the identity matrix of* H^d.

We must now prove the iterative estimate for the split procedure.

PROOF

To obtain an energy estimate for the scheme, we multiply by a test function $\partial_t^0 U_{j,k}^n$ and summarize over each time level.

The following result is given for the discretized equations; see also Equation (4.39).

$$(\mathcal{I} - \tau_n^2 \mathcal{N}_{4,\theta}) \partial_t^+ \partial_t^- U_{j,k}^n$$
$$- \mathcal{M}_4(\theta U_{j,k}^{n+1} + (1 - 2\theta) U_{j,k}^n + \theta U_{j,k}^{n-1}) = (0,0)^T. \qquad (4.45)$$

For $n \ge 1$ we can now rewrite Equation (4.45), and we multiply by a test function $\partial_t^0 U_{j,k}^n$. For the stability proof, we obtain

$$((\mathcal{I} - \tau_n^2 \mathcal{N}_{4,\theta}) \partial_t^+ \partial_t^- U_{j,k}^n, \partial_t^0 U_{j,k}^n)$$
$$- (\mathcal{M}_4(\theta U_{j,k}^{n+1} + (1 - 2\theta) U_{j,k}^n + \theta U_{j,k}^{n-1}), \partial_t^0 U_{j,k}^n) = 0. \qquad (4.46)$$

Multiplying by τ_n and summarizing over the time levels, we obtain

$$\sum_{n=1}^N ((\mathcal{I} - \tau_n^2 \mathcal{N}_{4,\theta}) \partial_t^+ \partial_t^- U_{j,k}^n, \partial_t^0 U_{j,k}^n) \tau_n$$

$$- \sum_{n=1}^N (\mathcal{M}_4(\theta U_{j,k}^{n+1} + (1 - 2\theta) U_{j,k}^n + \theta U_{j,k}^{n-1}), \partial_t^0 U_{j,k}^n) \tau_n = 0. \qquad (4.47)$$

For each term of the sum we can derive the identities. For example, for

$\mathcal{I} - \tau_n^2 \mathcal{N}_{4,\theta}$ we derive

$$((\mathcal{I} - \tau_n^2 \mathcal{N}_{4,\theta}) \partial_t^+ \partial_t^- U_{j,k}^n, \partial_t^0 U_{j,k}^n)_{\mathcal{T}_n}$$

$$= \frac{1}{2}((\mathcal{I} - \tau_n^2 \mathcal{N}_{4,\theta})(\partial_t^+ - \partial_t^-)U_{j,k}^n, (\partial_t^+ + \partial_t^-)U_{j,k}^n)$$

$$= \int_\Omega ((\mathcal{I} - \tau_n^2 \mathcal{N}_{4,\theta})(\partial_t^+ - \partial_t^-))^T (\partial_t^+ + \partial_t^-)U_{j,k}^n \, dx$$

$$\leq (1 - \tau_n^2 \tilde{\omega}) \int_\Omega (\partial_t^+ U_{j,k}^n)^2 (\partial_t^- U_{j,k}^n)^2 \, dx, \tag{4.48}$$

where the operator $\mathcal{I} - \tau_n^2 \mathcal{N}_{4,\theta}$ is symmetric and positive definite, and we can apply the weighted norm, see Remark 4.7 and [65].

We obtain the following result:

$$(1 - \tau_n^2 \tilde{\omega}) \int_\Omega (\partial_t^+ U_{j,k}^n)^2 (\partial_t^- U_{j,k}^n)^2 \, dx \tag{4.49}$$

$$= 1/2 \|(1 - \tau_n^2 \tilde{\omega})^{1/2} \partial_t^+ U_{j,k}^n\|^2$$

$$- 1/2 \|(1 - \tau_n^2 \tilde{\omega})^{1/2} \partial_t^- U_{j,k}^n\|^2. \tag{4.50}$$

For $-\mathcal{M}_4$, we find

$$(-\mathcal{M}_4(\theta U_{j,k}^{n+1} - (1 - 2\theta)U_{j,k}^n + \theta U_{j,k}^{n-1}), \partial_t^0 U_{j,k}^n)_{\mathcal{T}_n}$$

$$= 1/2(\mathcal{P}^+(U_{j,k}^n, \theta) - \mathcal{P}^-(U_{j,k}^n, \theta)). \tag{4.51}$$

As a result of the operators $\mathcal{P}^-(U_{j,k}^n, \theta) = \mathcal{P}^+(U_{j,k}^{n-1}, \theta)$ and $\partial_t^- U_{j,k}^n = \partial_t^+ U_{j,k}^{n-1}$, we can recursively derive the following result:

$$\|(1 - \tau_n^2 \tilde{\omega})^{1/2} \partial_t^+ U_{j,k}^N\|^2 + \mathcal{P}^+(U_{j,k}^N, \theta) \leq \|(1 - \tau_n^2 \tilde{\omega})^{1/2} \partial_t^+ U_{j,k}^0\|^2$$

$$+ \mathcal{P}^+(U_{j,k}^0, \theta), \tag{4.52}$$

where for $\theta \in [0.25, 0.5]$ we find $\mathcal{P}^+(U_{j,k}^N, \theta) \geq 0$ for all $N \in \mathbb{N}^+$, and, therefore, we have the unconditional stability. The scalar proof is also presented in the work of [133].

⬚

REMARK 4.8 For $\theta = \frac{1}{12}$, the splitting method is fourth-order accurate in time and space; see the following theorem.

⬚

THEOREM 4.6

We obtain a fourth-order accurate scheme in time and space for the splitting method, see (4.35)–(4.37), when $\theta = 1/12$. We find

$$\partial_t^+ \partial_t^- U_{j,k}^n - 1/12 \mathcal{M}_4(U_{j,k}^{n+1} - 2U_{j,k}^n + U_{j,k}^{n-1}) + \mathcal{M}_4 U_{j,k}^n$$

$$+ \mathcal{N}_{4,\theta}(U_{j,k}^{n+1} - 2U_{j,k}^n + U_{j,k}^{n-1}) = (0,0)^T, \tag{4.53}$$

where $\mathcal{M}_4$ is a fourth-order discretization scheme in space.

PROOF

We consider the following Taylor expansion:

$$\partial_{tt} U_{j,k}^n = \partial_t^+ \partial_t^- U_{j,k}^n - \frac{\tau_n^2}{12} \partial_{tttt} U_{j,k}^n + \mathcal{O}(\tau_n^4). \tag{4.54}$$

Furthermore, we have

$$\partial_{tttt} U_{j,k}^n \approx \mathcal{M}_4 \partial_{tt} U_{j,k}^n, \tag{4.55}$$

and we can rewrite (4.54) as

$$\partial_{tt} U_{j,k}^n \approx \partial_t^+ \partial_t^- U_{j,k}^n - \frac{\tau_n^2}{12} \mathcal{M}_4 \partial_{tt} U_{j,k}^n + \mathcal{O}(\tau_n^4)$$

$$\approx \partial_t^+ \partial_t^- U_{j,k}^n - \frac{1}{12} \mathcal{M}_4 (U_{j,k}^{n+1} - 2U_{j,k}^n + U_{j,k}^{n-1}) + \mathcal{O}(\tau_n^4). \tag{4.56}$$

The fourth-order time-stepping algorithm can be formulated as

$$\partial_t^+ \partial_t^- U_{j,k}^n - \frac{1}{12} \mathcal{M}_4 (U_{j,k}^{n+1} - 2U_{j,k}^n + U_{j,k}^{n-1}) - \mathcal{M}_4 U_{j,k}^n = (0,0)^T. \tag{4.57}$$

The splitting method, (4.35)–(4.37), becomes

$$\partial_t^+ \partial_t^- U_{j,k}^n - \frac{1}{12} \mathcal{M}_4 (U_{j,k}^{n+1} - 2U_{j,k}^n + U_{j,k}^{n-1}) - \mathcal{M}_4 U_{j,k}^n$$

$$- \mathcal{N}_{4, \frac{1}{12}} (U_{j,k}^{n+1} - 2U_{j,k}^n + U_{j,k}^{n-1}) = (0,0)^T, \tag{4.58}$$

and we obtain a fourth-order split scheme, cf. the scalar case [133].

□

REMARK 4.9 From Theorem 4.6 we find that the splitting method is fourth order in time for $\theta = 1/12$. For the stability analysis, the method is conditionally stable for $\theta \in (0, 0.25)$. So the splitting method will not restrict our stability condition for the fourth-order method with $\theta = 1/12$. □

To improve the time-step behavior, we introduce the iterative operator-splitting methods for hyperbolic equations.

4.3 Iterative Operator-Splitting Methods for Wave Equations

In the following, we present the iterative operator-splitting method as an extension of the traditional splitting method for wave equations.

Our objective is to repeat the splitting steps with the improved computed solutions. We solve a fixed-point iteration and we obtain higher-order results.

The iterative splitting method is given in the continuous formulation

$$\frac{d^2 c_i(t)}{dt^2} = Ac_i(t) + Bc_{i-1}(t) + f(t), \quad t \in [t^n, t^{n+1}],$$

$$\text{with } c_i(t^n) = c_{\text{sp}}^n, \quad \frac{dc_i(t^n)}{dt} = \frac{dc_{\text{sp}}^n}{dt}, \tag{4.59}$$

$$\frac{d^2 c_{i+1}(t)}{dt^2} = Ac_i(t) + Bc_{i+1}(t) + f(t), \quad t \in [t^n, t^{n+1}],$$

$$\text{with } c_{i+1}(t^n) = c_{\text{sp}}^n, \quad \frac{dc_{i+1}(t^n)}{dt} = \frac{dc_{\text{sp}}^n}{dt}, \tag{4.60}$$

where $c_0(t)$, $\frac{dc_0(t)}{dt}$ are fixed functions for each iteration. (Here, as before, c_{sp}^n, $\frac{dc_{\text{sp}}^n}{dt}$ denote known split approximations at the time level $t = t^n$.) The time-step is given as $\tau = t^{n+1} - t^n$. The split approximation at the time level $t = t^{n+1}$ is $c_{\text{sp}}^{n+1} = c_{2m+1}(t^{n+1})$.

For the discrete version of the iterative operator-splitting method, we apply the second-order discretization of the time derivations and obtain

$$c_i - 2c(t^n) + c(t^{n-1}) = \tau_n^2 A(\eta c_i + (1 - 2\eta)c(t^n) + \eta c(t^{n-1})) \tag{4.61}$$
$$+ \tau_n^2 B(\eta c_{i-1} + (1 - 2\eta)c(t^n) + \eta c(t^{n-1}))$$
$$+ \tau_n^2 (\eta f(t^{n+1}) + (1 - 2\eta)f(t^n) + \eta f(t^{n-1})),$$

$$c_{i+1} - 2c(t^n) + c(t^{n-1}) = \tau_n^2 A(\eta c_i + (1 - 2\eta)c(t^n) + \eta c(t^{n-1})) \tag{4.62}$$
$$+ \tau_n^2 B(\eta c_{i+1} + (1 - 2\eta)c(t^n) + \eta c(t^{n-1}))$$
$$+ \tau_n^2 (\eta f(t^{n+1}) + (1 - 2\eta)f(t^n) + \eta f(t^{n-1})),$$

where we iterate for $i = 1, 3, 5, \ldots$ and the starting solution $c_0(t)$, $\frac{dc_0(t)}{dt}$ are any fixed function for each iteration, for example, $c_0(t) = \frac{dc_0(t)}{dt} = 0$. The result is given as $c(t^{n+1})$ with the initial conditions $c(t^n) = c_{\text{sp}}^n$ and $\frac{dc(t^n)}{dt} = \frac{dc_{\text{sp}}^n}{dt}$, and $\eta \in [0, 0.5]$, using the fully coupled method for $\eta = 0$ and the decoupled method for $0 < \eta \leq 0.5$, which is a mixing of explicit and implicit Euler methods.

REMARK 4.10 The stop criterion is given as $|c_{k+1} - c_k| \leq \epsilon$, where $k \in 1, 3, 5, \ldots$ and $\epsilon \in \mathbb{R}^+$.

Therefore, the solution is given as $c(t^{n+1}) \approx c_{k+1}(t^{n+1}) = c_{\text{sp}}^{n+1}$. ☐

For the stability and consistency, we can rewrite Equations (4.61)–(4.62) in the continuous form in the operator equation as

$$\frac{d^2 C_i}{dt^2} = AC_i + \mathcal{F}_i, \tag{4.63}$$

where $C_i = (c_i, c_{i+1})^T$, and the operators are given as

$$A = \begin{bmatrix} A & 0 \\ A & B \end{bmatrix}, \quad \mathcal{F}_i = \begin{bmatrix} Bc_{i-1} \\ 0 \end{bmatrix}. \tag{4.64}$$

This equation is discussed in the following cases, in which the results are given with respect to stability and consistency.

4.3.1 Stability and Consistency Theory for the Iterative Operator-Splitting Method

The stability and consistency results can be obtained as we saw for the parabolic case. The operator equation with second-order time derivatives can be reformulated as a system of first-order time derivatives.

4.3.1.1 Consistency for the Iterative Operator-Splitting Method for Wave Equations

We now analyze the consistency and order of the local splitting error for the linear bounded operators $A, B : \mathbf{X} \to \mathbf{X}$, where $\mathbf{X}$ is a Banach space, see [200].

We assume our Cauchy problem for two linear operators for the second-order time derivative:

$$\frac{d^2 c}{dt^2} - Ac - Bc = 0, \text{ for } t \in (0, T), \tag{4.65}$$

$$\text{with } c(0) = c_0, \quad \frac{dc(0)}{dt} = c_1, \tag{4.66}$$

where c_0 and c_1 are the initial values, see Section 2.2.3.

We rewrite to a system of first-order time derivatives:

$$\partial_t c_1 - c_2 = 0, \text{ in } (0, T), \tag{4.67}$$

$$\partial_t c_2 - Ac_1 - Bc_1 = 0, \text{ in } (0, T), \tag{4.68}$$

$$\text{with } c_1(0) = c_0, c_2(0) = c_1, \tag{4.69}$$

where $c_0 = c(0)$ and $c_1 = \frac{dc(0)}{dt}$ are the initial values.

The iterative operator-splitting method (4.59)–(4.60) is rewritten to a system of splitting methods.

The method is given as

$$\partial_t c_{1,i} = c_{2,i}, \tag{4.70}$$

$$\partial_t c_{2,i} = Ac_{1,i} + Bc_{1,i-1}, \tag{4.71}$$

$$\text{with } c_{1,i}(t^n) = c_1(t^n), \ c_{2,i}(t^n) = c_2(t^n),$$

$$\partial_t c_{1,i+1} = c_{2,i+1}, \tag{4.72}$$

$$\partial_t c_{2,i+1} = Ac_{1,i} + Bc_{1,i+1}, \tag{4.73}$$

$$\text{with } c_{1,i+1}(t^n) = c_1(t^n), \ c_{2,i+1}(t^n) = c_2(t^n).$$

We start with $i = 1, 3, 5, \ldots, 2m + 1$ and the initial conditions $c_1(t^n)$, $c_2(t^n)$ are the approximate solutions at $t = t^n$.

We can obtain consistency with the underlying fundamental solution of the equation system.

THEOREM 4.7

Let $A, B \in \mathcal{L}(\mathbf{X})$ be given linear bounded operators. Then the abstract Cauchy problem (4.65)–(4.66) has a unique solution, and the iterative splitting method (4.70)–(4.73) for $i = 1, 3, \ldots, 2m + 1$ is consistent with the order of the consistency $\mathcal{O}(\tau_n^{2m})$.

The error estimate is given as

$$\|e_i\| = K\|B\|\tau_n\|e_{i-1}\| + \mathcal{O}(\tau_n^2), \tag{4.74}$$

where $e_i = \max\{|e_{1,i}|, |e_{i,2}|\}$.

PROOF

We derive the underlying consistency of the operator-splitting method.

Let us consider the iteration (4.61)–(4.62) on the subinterval $[t^n, t^{n+1}]$. For the local error function $e_i(t) = c(t) - c_i(t)$, we have the relations

$$
\begin{aligned}
\partial_t e_{1,i}(t) &= e_{2,i}(t), & t &\in [t^n, t^{n+1}], \\
\partial_t e_{2,i}(t) &= A e_{1,i}(t) + B e_{1,i-1}(t), & t &\in [t^n, t^{n+1}], \\
\partial_t e_{1,i+1}(t) &= e_{2,i+1}(t), & t &\in [t^n, t^{n+1}], \\
\partial_t e_{2,i+1}(t) &= A e_{1,i}(t) + B e_{1,i+1}(t), & t &\in [t^n, t^{n+1}],
\end{aligned} \tag{4.75}
$$

for $i = 0, 2, 4, \ldots$, with $e_0(0) = 0$ and $e_{-1}(t) = c(t)$. We use the notation $\mathbf{X}^4 = \Pi_{i=1}^4 \mathbf{X}$ for the product space, which enables the norm $\|(u_1, u_2, u_3, u_4)^T\| = \max\{\|u_1\|, \|u_2\|, \|u_3\|, \|u_4\|\}$ with $(u_1, u_2, u_3, u_4 \in \mathbf{X})$, and for each component we assume the supremum norm.

The elements $\mathcal{E}_i(t)$, $\mathcal{F}_i(t) \in \mathbf{X}^4$ and the linear operator $\mathcal{A} : \mathbf{X}^4 \to \mathbf{X}^4$ are defined as

$$
\mathcal{E}_i(t) = \begin{bmatrix} e_{1,i}(t) \\ e_{2,i}(t) \\ e_{1,i+1}(t) \\ e_{2,i+1}(t) \end{bmatrix}, \quad
\mathcal{F}_i(t) = \begin{bmatrix} 0 \\ B e_{1,i-1}(t) \\ 0 \\ 0 \end{bmatrix}, \quad
\mathcal{A} = \begin{bmatrix} 0 & I & 0 & 0 \\ A & 0 & 0 & 0 \\ 0 & I & 0 & I \\ A & 0 & B & 0 \end{bmatrix}. \tag{4.76}
$$

Then, using the notations (4.76), the relations of (4.75) can be written:

$$
\begin{aligned}
\partial_{tt} \mathcal{E}_i(t) &= \mathcal{A} \mathcal{E}_i(t) + \mathcal{F}_i(t), \quad t \in (t^n, t^{n+1}], \\
\mathcal{E}_i(t^n) &= 0.
\end{aligned} \tag{4.77}
$$

Due to our assumptions, $\mathcal{A}$ is a generator of the one-parameter C_0 semigroup $(\exp \mathcal{A}t)_{t \geq 0}$; hence, using the variations of constants formula, the solution to the abstract Cauchy problem (4.77) with homogeneous initial conditions can be written as

$$\mathcal{E}_i(t) = c_0 \int_{t^n}^{t} \exp(\mathcal{A}(t-s))\mathcal{F}_i(s)ds, \tag{4.78}$$

with $t \in [t^n, t^{n+1}]$. (See, e.g., [63].) Hence, using the denotation

$$\|\mathcal{E}_i\|_{\infty} = \sup_{t \in [t^n, t^{n+1}]} \|\mathcal{E}_i(t)\|, \tag{4.79}$$

we have

$$\|\mathcal{E}_i\|(t) \leq \|\mathcal{F}_i\|_{\infty} \int_{t^n}^{t} \|\exp(\mathcal{A}(t-s))\|ds$$

$$= \|B\|\|e_{1,i-1}\| \int_{t^n}^{t} \|\exp(\mathcal{A}(t-s))\|ds, \quad t \in [t^n, t^{n+1}]. \tag{4.80}$$

Because $(\mathcal{A}(t))_{t \geq 0}$ is a semigroup, the *growth estimation*

$$\|\exp(\mathcal{A}t)\| \leq K \exp(\omega t), \quad t \geq 0, \tag{4.81}$$

holds for some numbers $K \geq 0$ and $\omega \in \mathbb{R}$, cf. [63].

The estimations (4.80) and (4.81) result in

$$\|\mathcal{E}_i\|_{\infty} = K\|B\|\tau_n\|e_{i-1}\| + \mathcal{O}(\tau_n^2). \tag{4.82}$$

where $\|e_{i-1}\| = \max\{\|e_{1,i-1}\|, \|e_{2,i-1}\|\}$.

Taking into account the definition of $\mathcal{E}_i$ and the norm $\|\cdot\|_{\infty}$, we obtain

$$\|e_i\| = K\|B\|\tau_n\|e_{i-1}\| + \mathcal{O}(\tau_n^2), \tag{4.83}$$

and hence,

$$\|e_{i+1}\| = K_1\tau_n^2\|e_{i-1}\| + \mathcal{O}(\tau_n^3), \tag{4.84}$$

which proves our statement.

□

REMARK 4.11 The proof is aligned to scalar temporal first-order derivatives, see [70]. The generalization can also be extended to higher-order hyperbolic equations, which are reformulated in first-order systems. □

4.3.1.2 Stability for the Iterative Operator-Splitting Method for Wave Equations

The following stability theorem is given for the wave equation performed with the iterative splitting method, see (4.70)–(4.73).

The convergence is examined in a general Banach space setting, and we can prove the following stability theorem.

THEOREM 4.8

Let us consider the system of linear differential equations used for the spatial discretized wave equation:

$$\frac{dc_1}{dt} = c_2, \tag{4.85}$$

$$\frac{dc_2}{dt} = Ac_1 + Bc_1, \tag{4.86}$$

$$\text{with } c_1(t^n) = c(t^n), \ c_2(t^n) = \frac{dc(t^n)}{dt},$$

where the operators $A, B : \mathbf{X} \to \mathbf{X}$ are linear and densely defined in the real Banach space $\mathbf{X}$, see [201] and Section 3.2.4. We can define a norm on the product space $\mathbf{X} \times \mathbf{X}$ with $||(u, v)^T|| = \max\{||u||, ||v||\}$.

We rewrite the equations (4.85)–(4.86) and obtain

$$\frac{d\tilde{c}(t)}{dt} = \tilde{A}\tilde{c}(t) + \tilde{B}\tilde{c}(t),$$
$$\tilde{c}(t^n) = \tilde{c}^n, \tag{4.87}$$

where $\tilde{c}^n = (c(t^n), \frac{dc(t^n)}{dt})^T$, $\tilde{A} = \begin{pmatrix} 0 & 1/2I \\ A & 0 \end{pmatrix}$ and $\tilde{B} = \begin{pmatrix} 0 & 1/2I \\ B & 0 \end{pmatrix}$.

We assume that $\tilde{A}, \tilde{B} : \mathbf{X}^2 \to \mathbf{X}^2$ are given linear bounded operators that generate the C_0 semigroup, and $\tilde{c}, \tilde{c}^n \in \mathbf{X}^2$ are given elements.

We also assume $\lambda_{\tilde{A}}$ is an eigenvalue of $\tilde{A}$ and $\lambda_{\tilde{B}}$ is an eigenvalue of $\tilde{B}$.

Then the linear iterative operator-splitting method for wave equations (4.70)–(4.73) is stable with the following result:

$$||\tilde{c}_{i+1}(t^{n+1})|| \leq \tilde{K} \sum_{j=0}^{i+1} ||\tilde{c}^n|| \tau^j \lambda_{\max}^j, \tag{4.88}$$

where $\tilde{K} > 0$ is a constant and $\tilde{c}^n = (c(t^n), \frac{dc(t^n)}{dt})^T$ is the initial condition, $\tau = (t^{n+1} - t^n)$ is the time-step, and $\lambda_{\max}$ is the maximal eigenvalue of the linear and bounded operators $\tilde{A}$ and $\tilde{B}$.

We discuss Theorem 4.8 in the following proof.

PROOF

Let us consider the iteration (4.70)–(4.73) on the subinterval $[t^n, t^{n+1}]$.

Then we obtain the eigenvalues of the following linear and bounded operators. Because of the well-posed problem, we have $\lambda_{\tilde{A}}$ eigenvalue of $\tilde{A}$, $\lambda_{\tilde{B}}$ eigenvalue of $\tilde{B}$, see [125] and [201].

Then our iteration methods are given with the eigenvalues as follows:

$$\partial_t \tilde{c}_i(t) = \lambda_{\tilde{A}} \tilde{c}_i(t)) + \lambda_{\tilde{B}} \tilde{c}_{i-1}(t), \quad t \in (t^n, t^{n+1}],$$
$$\tilde{c}_i(t^n) = \tilde{c}^n, \tag{4.89}$$

and

$$\partial_t \tilde{c}_{i+1}(t) = \lambda_{\tilde{A}} \tilde{c}_i(t) + \lambda_{\tilde{B}} \tilde{c}_{i+1}(t), \quad t \in (t^n, t^{n+1}],$$
$$\tilde{c}_{i+1}(t^n) = \tilde{c}^n, \tag{4.90}$$

for $i = 1, 3, 5, \ldots$, with $\tilde{c}^n = (c(t^n), \frac{dc(t^n)}{dt})^T$.

The equations can be estimated with

$$\tilde{c}_i(t^{n+1}) = \exp(\lambda_{\tilde{A}} \tau) \tilde{c}^n + \int_{t^n}^{t^{n+1}} \exp(\lambda_{\tilde{A}}(t^{n+1} - s)) \lambda_{\tilde{B}} \tilde{c}_{i-1}(s) ds, \tag{4.91}$$

where we can estimate the result as

$$||\tilde{c}_i(t^{n+1})|| \le K_1 ||\tilde{c}^n|| + \tau K_2 \lambda_{\tilde{B}} ||\tilde{c}_{i-1}(t^{n+1})||. \tag{4.92}$$

where K_1, K_2 are the growth estimates for the exp-functions, see Section 4.3.1.1, and we neglect the higher-order terms.

We can also estimate the second equation:

$$\tilde{c}_{i+1}(t^{n+1}) = \exp(\lambda_{\tilde{B}} \tau) \tilde{c}^n + \int_{t^n}^{t^{n+1}} \exp(\lambda_{\tilde{B}}(t^{n+1} - s)) \lambda_{\tilde{A}} \tilde{c}_i(s) ds, \tag{4.93}$$

which can be estimated as

$$||\tilde{c}_{i+1}(t^{n+1})|| \le K_3 ||\tilde{c}^n|| + \tau K_4 \lambda_{\tilde{A}} ||\tilde{c}_i(t^{n+1})||. \tag{4.94}$$

where K_3, K_4 are the growth estimates for the exp-functions, see Section 4.3.1.1, and we neglect the higher-order terms.

With the recursive argument and the maximum of the eigenvalues, we can estimate the equations:

$$||\tilde{c}_{i+1}(t^{n+1})|| \le ||\tilde{c}^n|| \sum_{j=0}^{i+1} K_j \tau^j \lambda_{\max}^j, \tag{4.95}$$

$$||\tilde{c}_{i+1}(t^{n+1})|| \le \tilde{K} ||\tilde{c}^n|| \sum_{j=0}^{i+1} \tau^j \lambda_{\max}^j, \tag{4.96}$$

where $\tilde{K}$ is the maximum of all constants and $\lambda_{\max} = \max\{\lambda_{\tilde{A}}, \lambda_{\tilde{B}}\}$. ☐

REMARK 4.12 We have stability for sufficient small time-steps τ. Based on the estimation with the eigenvalues, we can do the same technique for unbounded operators that are boundable in a sector. In addition, accurate estimates can be derived using the techniques of the mild or weak solutions, see [201]. ☐

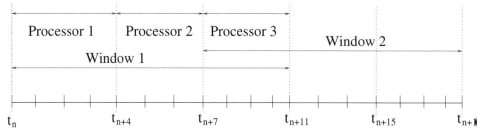

FIGURE 4.2: Parallelization of the time intervals.

4.4 Parallelization of Time Decomposition Methods

The parallelization of the splitting methods is suited in applications for CFD or convection-diffusion-reaction processes and large-scale computations. We present two possible parallelization techniques:

1) Windowing (see [194]);

2) Block-wise clustering (see [35] and [36]).

4.4.1 Windowing

The first direction is the parallelization of the time-stepping process. A large time-step is decoupled into smaller time-steps and is solved independently with higher-order time discretization methods by any processor. The core concept of parallelization is windowing, in which the processor has one or more time-steps to compute and shares the end result of the computation as an initial condition for the next processor.

A graphical visualization of such a parallelization technique is presented in Figure 4.2.

REMARK 4.13 The windowing technique is efficient for our iterative splitting methods, because they can be parallelized on the global level. Therefore, in each local time partition, the processors solve the local operator equation and the communication is done on the global level to forward the initial solutions for the next time partition. ▯

4.4.2 Block-Wise Clustering

A further parallelization of the methods is the block-wise decoupling of the matrix into simpler solvable partitions. In this technique, we parallelize on each operator level. Therefore, the parallelization is based on solving each submatrix independently and summarizing the results in an additional step to get the results.

The local operator equations are given as a system of first-order differential equations:

$$\frac{du}{dt} = Au, \text{ for } t \in [0, T],$$
$$A_1 + A_2 \approx A, \tag{4.97}$$
$$u(0) = u_0,$$

where the matrices are given as

$$A = \begin{pmatrix} A_{11} & A_{12} & \cdots & & A_{1n} \\ A_{21} & A_{22} & & & \vdots \\ \vdots & & \ddots & & A_{n-1n} \\ A_{n1} & \cdots & A_{nn-1} & & A_{nn} \end{pmatrix}, \tag{4.98}$$

and the entries are denoted as $A_{ii} \gg A_{ij}$, $\forall\, i, j = 1, \ldots, n$ and $i \neq j$ (i.e., diagonal dominant matrices).

We assume that the outer diagonal entries can be neglected, and the additive operators are given as

$$A_1 + A_2 = \begin{pmatrix} A_{11} & 0 & \cdots & 0 \\ 0 & A_{22} & & \vdots \\ \vdots & & \ddots & 0 \\ 0 & \cdots & 0 & A_{nn} \end{pmatrix}$$

$$= \begin{pmatrix} A_{11,1} & 0 & \cdots & 0 \\ 0 & A_{22,1} & & \vdots \\ \vdots & & \ddots & 0 \\ & \cdots & 0 & A_{nn,1} \end{pmatrix} + \begin{pmatrix} A_{11,2} & 0 & \cdots & 0 \\ 0 & A_{22,2} & & \vdots \\ \vdots & & \ddots & 0 \\ 0 & \cdots & 0 & A_{nn,2} \end{pmatrix}. \tag{4.99}$$

The block-wise clustering allows us to parallelize with respect to each independent block.

We can also implement our splitting method on the local operator level with A_1 and A_2.

REMARK 4.14 The block-wise clustering allows a parallelization on the local level. Therefore, the outer diagonal entries of the underlying matrices are negligible or very small. The processors can be distributed to the various blocks to compute the problems independently. These concepts are also discussed in [35] and [36]. □

4.5 Nonlinear Iterative Operator-Splitting Methods

In this chapter we deal with the nonlinear operators. Behind is the linearization with the Newton method. We could linearize the nonlinear operators and deal with linear methods.

4.5.1 Linearization with the Waveform Relaxation

We can formulate the operator-splitting methods into waveform-relaxation methods. This generalization helps us to unify the theory around the splitting methods.

$$F_i(t, u, v) = f_i(t, v_1, v_2, \ldots, v_{i-1}, u_i, \ldots, u_d)$$
$$+ \frac{\partial f_i}{\partial v_i}(t, v_1, \ldots, v_{i-1}, u_i, \ldots, u_d)'(v_i - u_i), \tag{4.100}$$

where the iteration is the Gauss-Seidel-waveform-relaxation Newton method, cf. [194].

For the two operator equation,

$$\partial_t c = A(c) + B(c) \tag{4.101}$$

we use the linearization

$$A(c^i) \approx A(c^{i-1}) + A'(c^{i-1})(c^i - c^{i-1}) \tag{4.102}$$

and obtain the following linearized equation:

$$\partial_t c^i = A(c^{i-1}) + A'(c^{i-1})(c^i - c^{i-1}) + B(c^{i-1}) + B'(c^{i-1})(c^i - c^{i-1}),$$

where i is the iteration index, A' the Jacobian matrix of A, and we start with the initial value $c^0 = c(t^n)$.

We can rewrite the equation system as a formal A-B splitting:

$$\partial_t c^{i,*} = A(c^{i-1,*}) + A'(c^{i-1,*})(c^{i,*} - c^{i-1,*}), \tag{4.103}$$
$$\text{with } c^{i,*}(t^n) = c(t^n),$$
$$\partial_t c^{i,**} = B(c^{i-1,**}) + B'(c^{i-1,**})(c^{i,**} - c^{i-1,**}), \tag{4.104}$$
$$\text{with } c^{i,**}(t^n) = c(t^n), .$$

where $i = 1, 2, \ldots, M$ and the stop criterion is given as $\min(|c^{i,*} - c^{i-1,*}|, |c^{i,**} - c^{i-1,**}|) = err$, and $err \in \mathbb{R}^+$.

REMARK 4.15 The nonlinear iterative operator-splitting method can be discussed with different linearization techniques, see [92]. The linearization techniques are taken into account Newton's method or fixpoint schemes, see more results in [93]. ▯

Chapter 5

Spatial Decomposition Methods

5.1 Domain Decomposition Methods Based on Iterative Operator-Splitting Methods

The combined spatiotemporal iterative operator-splitting method combines the Schwarz waveform-relaxation and the iterative operator-splitting method.

The following algorithm iterates with a fixed splitting discretization step size τ. On the time interval $[t^n, t^{n+1}]$, we solve the following subproblems consecutively for $i = 1, 3, \ldots 2m+1$ and $j = 1, 3, \ldots 2m+1$. In this notation, i represents the iteration index for the temporal splitting and j represents the iteration index for the spatial splitting.

$$\frac{dc_{i,j}(t)}{dt} = A|_{\overline{\Omega}_1} c_{i,j}(t) + A|_{\overline{\Omega}_2} c_{i,j-1}(t)$$
$$+ B|_{\overline{\Omega}_1} c_{i-1,j}(t) + B|_{\overline{\Omega}_2} c_{i-1,j-1}(t),$$
$$\text{with } c_{i,j}(t^n) = c^n, \tag{5.1}$$

$$\frac{dc_{i+1,j}(t)}{dt} = A|_{\overline{\Omega}_1} c_{i,j}(t) + A|_{\overline{\Omega}_2} c_{i,j-1}(t)$$
$$+ B|_{\overline{\Omega}_1} c_{i+1,j}(t) + B|_{\overline{\Omega}_2} c_{i-1,j-1}(t),$$
$$\text{with } c_{i+1,j}(t^n) = c^n, \tag{5.2}$$

$$\frac{dc_{i,j+1}(t)}{dt} = A|_{\overline{\Omega}_1} c_{i,j}(t) + A|_{\overline{\Omega}_2} c_{i,j+1}(t)$$
$$+ B|_{\overline{\Omega}_1} c_{i+1,j}(t) + B|_{\overline{\Omega}_2} c_{i-1,j-1}(t),$$
$$\text{with } c_{i,j+1}(t^n) = c^n, \tag{5.3}$$

$$\frac{dc_{i+1,j+1}(t)}{dt} = A|_{\overline{\Omega}_1} c_{i,j}(t) + A|_{\overline{\Omega}_2} c_{i,j+1}(t)$$
$$+ B|_{\overline{\Omega}_1} c_{i+1,j}(t) + B|_{\overline{\Omega}_2} c_{i+1,j+1}(t),$$
$$\text{with } c_{i+1,j+1}(t^n) = c^n, \tag{5.4}$$

where $c_{0,0}(t), c_{1,0}(t)$, and $c_{0,1}(t)$ are fixed functions, for example, $c_{0,0}(t) = c_{1,0}(t) = c_{0,1}(t) = 0$, for each iteration. c^n is the known split approximation at the time level $t = t^n$, cf. [70]. The boundary conditions are Neumann conditions that are embedded in the equations, see Chapter 2 and [178]. We have the domain Ω with $\Omega = \Omega_1 \cup \Omega_2$, $\Omega_1 \cap \Omega_2 = \Omega_{1,2}$ and the restriction to each operator that is, $A|_{\Omega_i}, B|_{\Omega_i}$ with $i = 1, 2$ for two subdomains. The coupling is done at the intermediate boundaries $\overline{\Omega}_1 \cap \overline{\Omega}_2 = \Gamma_{\Omega_1, \Omega_2}$ ($\Omega_{1,2} = \emptyset$, i.e. mass zero, see [179]) for the nonoverlapping method or at the overlapping set ($\Omega_{1,2} \neq \emptyset$, i.e., of mass 1, see [179]) for the overlapping method. The overlaps are presented in Figure 5.1.

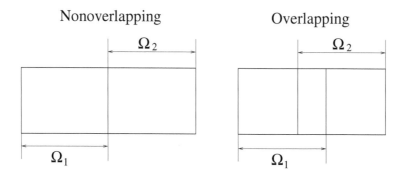

| Nonoverlapping | Overlapping |

FIGURE 5.1: Graphical visualization of the overlaps.

REMARK 5.1 We extend the splitting method with respect to the underlying spatial discretized operators. For each subdomain, we redefine the corresponding operators into the subdomain. ▯

5.1.1 The Nonoverlapping Spatiotemporal Iterative Splitting Method

For the semi-discretization in space, we introduce the variable k as the node for the point x_k, and we have $k \in \{0, \ldots, p\}$, where p is the number of nodes. We concentrate on the one-dimensional nonoverlapping case. For the nonoverlapping case we assume the decomposition in space, where Ω_1 consists of the points $0, \ldots, p/2$, and Ω_2 consists of $p/2 + 1, \ldots, p$, whereby we assume

p to be even. Thus, we assume $\Omega_1 \cap \Omega_2 = \emptyset$ leading to the following algorithm:

$$\frac{d(c_{i,j})_k(t)}{dt} = \tilde{A}|_{\overline{\Omega}_1}(c_{i,j})_k(t) + \tilde{A}|_{\overline{\Omega}_2}(c_{i,j-1})_k(t)$$
$$+ \tilde{B}|_{\overline{\Omega}_1}(c_{i-1,j})_k(t) + \tilde{B}|_{\overline{\Omega}_2}(c_{i-1,j-1})_k(t),$$
with $(c_{i,j})_k(t^n) = (c^n)_k,$ \hfill (5.5)

$$\frac{d(c_{i+1,j})_k(t)}{dt} = \tilde{A}|_{\overline{\Omega}_1}(c_{i,j})_k(t) + \tilde{A}|_{\overline{\Omega}_2}(c_{i,j-1})_k(t)$$
$$+ \tilde{B}|_{\overline{\Omega}_1}(c_{i+1,j})_k(t) + \tilde{B}|_{\overline{\Omega}_2}(c_{i-1,j-1})_k(t),$$
with $(c_{i+1,j})_k(t^n) = (c^n)_k,$ \hfill (5.6)

$$\frac{d(c_{i,j+1})_k(t)}{dt} = \tilde{A}|_{\overline{\Omega}_1}(c_{i,j})_k(t) + \tilde{A}|_{\overline{\Omega}_2}(c_{i,j+1})_k(t)$$
$$+ \tilde{B}|_{\overline{\Omega}_1}(c_{i+1,j})_k(t) + \tilde{B}|_{\overline{\Omega}_2}(c_{i-1,j-1})_k(t),$$
with $(c_{i,j+1})_k(t^n) = (c^n)_k,$ \hfill (5.7)

$$\frac{d(c_{i+1,j+1})_k(t)}{dt} = \tilde{A}|_{\overline{\Omega}_1}(c_{i,j})_k(t) + \tilde{A}|_{\overline{\Omega}_2}(c_{i,j+1})_k(t)$$
$$+ \tilde{B}|_{\overline{\Omega}_1}(c_{i+1,j})_k(t) + \tilde{B}|_{\overline{\Omega}_2}(c_{i+1,j+1})_k(t),$$
with $(c_{i+1,j+1})_k(t^n) = (c^n)_k,$ \hfill (5.8)

where c^n is the known split approximation at the time level $t = t^n$, cf. [70].

For the one-dimensional problem, the operators in the equations with the variable $(c_{i,j})_k$ above are given as

$$\tilde{A}|_{\overline{\Omega}_1}(c_{i,j})_k = \begin{cases} Ac_{i,j} &, \text{for } k \in \{0, \ldots, p/2\} \\ 0 &, \text{for } k \in \{p/2 + 1, \ldots, p\} \end{cases}, \hfill (5.9)$$

$$\tilde{A}|_{\overline{\Omega}_2}(c_{i,j})_k = \begin{cases} 0 &, \text{for } k \in \{0, \ldots, p/2 - 1\} \\ Ac_{i,j} &, \text{for } k \in \{p/2, \ldots, p\} \end{cases}, \hfill (5.10)$$

where $p/2$ is the intermediate point at the overlap.

The assignments for the operator B are similar:

$$\tilde{B}|_{\overline{\Omega}_1}(c_{i,j})_k = \begin{cases} Bc_{i,j} &, \text{for } k \in \{0, \ldots, p/2\} \\ 0 &, \text{for } k \in \{p/2 + 1, \ldots, p\} \end{cases}, \hfill (5.11)$$

$$\tilde{B}|_{\overline{\Omega}_2}(c_{i,j})_k = \begin{cases} 0 &, \text{for } k \in \{0, \ldots, p/2 - 1\} \\ Bc_{i,j} &, \text{for } k \in \{p/2, \ldots, p\} \end{cases}, \hfill (5.12)$$

where $p/2$ is the intermediate point at the overlap.

In the same manner as above, the other operators in (5.5)–(5.8) can also be defined.

REMARK 5.2 We can also generalize our results for the multidimensional case. In that case, we have a set of intermediate boundary nodes, which couple the multidimensional domains, see [179].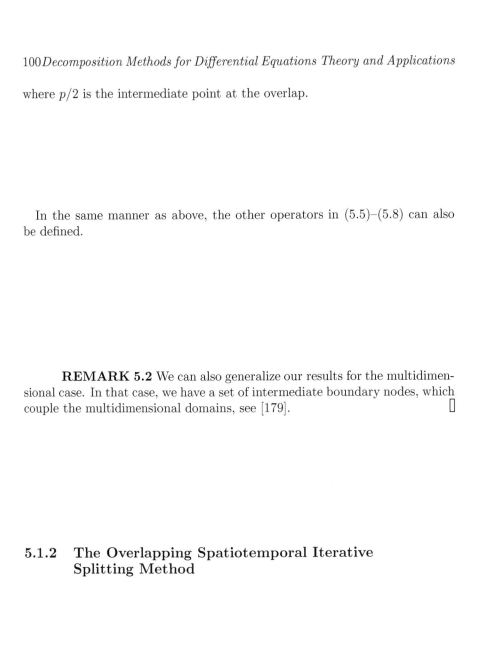

5.1.2 The Overlapping Spatiotemporal Iterative Splitting Method

We introduce for the semi-discretization in space the variable k as the node for the point x_k, and we have $k \in \{0, \ldots, p\}$, where p is the number of nodes. We concentrate on the one-dimensional overlapping case. For the overlapping case we assume $\Omega_1 \cap \Omega_2 \neq \emptyset$. We have the following sets: $\overline{\Omega}\backslash\Omega_2 = \{0, \ldots, p_1\}$, $\Omega_1 \cap \Omega_2 = \{p_1 + 1, \ldots, p_2 - 1\}$ and $\overline{\Omega}\backslash\Omega_1 = \{p_2, \ldots, p\}$. We assume that the nodes of the intermediate boundaries are in the sets $\overline{\Omega}\backslash\Omega_2$ and $\overline{\Omega}\backslash\Omega_1$, and we can also add these points into the overlapping domain.

Assuming $p_1 < p_2 < p$, we can derive the following overlapping algorithm:

$$\frac{d(c_{i,j})_k(t)}{dt} = \tilde{A}|_{\overline{\Omega}\backslash\Omega_2}(c_{i,j})_k(t) + \tilde{A}|_{\Omega_1\cap\Omega_2}(c_{i,j},c_{i,j-1})_k(t)$$
$$+ \tilde{A}|_{\overline{\Omega}\backslash\Omega_1}(c_{i,j-1})_k(t)$$
$$+ \tilde{B}|_{\overline{\Omega}\backslash\Omega_2}(c_{i-1,j})_k(t) + \tilde{B}|_{\Omega_1\cap\Omega_2}(c_{i-1,j},c_{i-1,j-1})_k(t)$$
$$+ \tilde{B}|_{\overline{\Omega}\backslash\Omega_1}(c_{i-1,j-1})_k(t),$$
$$\text{with } (c_{i,j})_k(t^n) = (c^n)_k, \tag{5.13}$$

$$\frac{d(c_{i+1,j})_k(t)}{dt} = \tilde{A}|_{\overline{\Omega}\backslash\Omega_2}(c_{i,j})_k(t) + \tilde{A}|_{\Omega_1\cap\Omega_2}(c_{i,j},c_{i,j-1})_k(t)$$
$$+ \tilde{A}|_{\overline{\Omega}\backslash\Omega_1}(c_{i,j-1})_k(t)$$
$$+ \tilde{B}|_{\overline{\Omega}\backslash\Omega_2}(c_{i+1,j})_k(t) + \tilde{B}|_{\Omega_1\cap\Omega_2}(c_{i+1,j},c_{i-1,j-1})_k(t)$$
$$+ \tilde{B}|_{\overline{\Omega}\backslash\Omega_1}(c_{i-1,j-1})_k(t),$$
$$\text{with } (c_{i+1,j})_k(t^n) = (c^n)_k, \tag{5.14}$$

$$\frac{d(c_{i,j+1})_k(t)}{dt} = \tilde{A}|_{\overline{\Omega}\backslash\Omega_2}(c_{i,j})_k(t) + \tilde{A}|_{\Omega_1\cap\Omega_2}(c_{i,j+1},c_{i,j})_k(t)$$
$$+ \tilde{A}|_{\overline{\Omega}\backslash\Omega_1}(c_{i,j+1})_k(t)$$
$$+ \tilde{B}|_{\overline{\Omega}\backslash\Omega_2}(c_{i+1,j})_k(t) + \tilde{B}|_{\Omega_1\cap\Omega_2}(c_{i+1,j},c_{i-1,j-1})_k(t)$$
$$+ \tilde{B}|_{\overline{\Omega}\backslash\Omega_1}(c_{i-1,j-1})_k(t),$$
$$\text{with } (c_{i,j+1})_k(t^n) = (c^n)_k, \tag{5.15}$$

$$\frac{d(c_{i+1,j+1})_k(t)}{dt} = \tilde{A}|_{\overline{\Omega}\backslash\Omega_2}(c_{i,j})_k(t) + \tilde{A}|_{\Omega_1\cap\Omega_2}(c_{i,j+1},c_{i,j})_k(t)$$
$$+ \tilde{A}|_{\overline{\Omega}\backslash\Omega_1}(c_{i,j+1})_k(t)$$
$$+ \tilde{B}|_{\overline{\Omega}\backslash\Omega_2}(c_{i+1,j})_k(t) + \tilde{B}|_{\Omega_1\cap\Omega_2}(c_{i+1,j},c_{i+1,j+1})_k(t)$$
$$+ \tilde{B}|_{\overline{\Omega}\backslash\Omega_1}(c_{i+1,j+1})_k(t),$$
$$\text{with } (c_{i+1,j+1})_k(t^n) = (c^n)_k, \tag{5.16}$$

where c^n is the known split approximation at the time level $t = t^n$, cf. [70].

We have the following operators:

$$\tilde{A}|_{\overline{\Omega}\backslash\Omega_2}(c_{i,j})_k = \begin{cases} A(c_{i,j})_k & , \text{for } k \in \{0,\ldots,p_1\} \\ 0 & , \text{for } k \in \{p_1+1,\ldots,p\} \end{cases}, \tag{5.17}$$

$$\tilde{A}|_{\Omega_1 \cap \Omega_2}(c_{i,j}, c_{i,j+1})_k = \begin{cases} 0 & \text{, for } k \in \{0, \dots, p_1 - 1\} \\ A((c_{i,j} + c_{i,j+1})/2)_k & \text{, for } k \in \{p_1, \dots, p_2\} \\ 0 & \text{, for } k \in \{p_2 + 1, \dots, p\} \end{cases},$$

$$(5.18)$$

$$\tilde{A}|_{\overline{\Omega} \backslash \Omega_1}(c_{i,j})_k = \begin{cases} 0 & \text{, for } k \in \{0, \dots, p_2 - 1\} \\ A(c_{i,j})_k & \text{, for } k \in \{p_2, \dots, p\} \end{cases}. \qquad (5.19)$$

The assignments are similar for the operator B.

$$\tilde{B}|_{\overline{\Omega} \backslash \Omega_2}(c_{i,j})_k = \begin{cases} B(c_{i,j})_k & \text{, for } k \in \{0, \dots, p_1\} \\ 0 & \text{, for } k \in \{p_1 + 1, \dots, p\} \end{cases}, \qquad (5.20)$$

$$\tilde{B}|_{\Omega_1 \cap \Omega_2}(c_{i,j}, c_{i,j+1})_k = \begin{cases} 0 & \text{, for } k \in \{0, \dots, p_1 - 1\} \\ B((c_{i,j} + c_{i,j+1})/2)_k & \text{, for } k \in \{p_1, \dots, p_2\} \\ 0 & \text{, for } k \in \{p_2 + 1, \dots, p\} \end{cases},$$

$$(5.21)$$

$$\tilde{B}|_{\overline{\Omega} \backslash \Omega_1}(c_{i,j})_k = \begin{cases} 0 & \text{, for } k \in \{0, \dots, p_2 - 1\} \\ B(c_{i,j})_k & \text{, for } k \in \{p_2, \dots, p\} \end{cases}. \qquad (5.22)$$

In the same manner as above, the other operators in (5.13)–(5.16) can be defined.

REMARK 5.3 We can generalize our results for the multidimensional case. The overlapping case is a set of the overlapping domain that couples the multidimensional domains, see [179]. □

5.1.3 Error Analysis and Convergence of the Combined Method

THEOREM 5.1
Let us consider the linear operator equation in a Banach space $\mathbf{X}$:

$$\frac{dc(t)}{dt} = A_1 c(t) + A_2 c(t) + B_1 c(t) + B_2 c(t), \quad 0 < t \le T,$$

$$c(0) = c_0,$$

where $A_1, A_2, B_1, B_2, A_1 + A_2 + B_1 + B_2 : \mathbf{X} \to \mathbf{X}$ *are given linear operators being generators of the* C_0 *semigroup and* $c_0 \in \mathbf{X}$ *is a given element. Then the iteration process* (5.1)–(5.4) *is convergent with a convergence rate of one.*

We obtain the iterative result: $\|e_{i,j}(t)\| \leq K\tau_n\|e_{i-1,j-1}(t)\| + \mathcal{O}(\tau_n^2)$, where $\tau_n = t^{n+1} - t^n$ and $t \in [t^n, t^{n+1}]$.

PROOF

Let us consider the iteration (5.1)–(5.4) on the subinterval $[t^n, t^{n+1}]$. We consider the case of the exact initial conditions given as $c_{i,j}(t^n) = c_{i+1,j}(t^n) = c_{i,j+1}(t^n) = c_{i+1,j+1}(t^n) = c(t^n)$, where a generalization is also possible. For the error functions $e_{i,j}(t) := c(t) - c_{i,j}(t)$, we have the relations

$$\frac{de_{i,j}(t)}{dt} = A_1\,e_{i,j}(t) + A_2\,e_{i,j-1}(t) + B_1\,e_{i-1,j}(t) + B_2\,e_{i-1,j-1}(t),$$
$$e_{i,j}(t^n) = 0, \tag{5.23}$$

$$\frac{de_{i+1,j}(t)}{dt} = A_1\,e_{i,j}(t) + A_2\,e_{i,j-1}(t) + B_1\,e_{i+1,j}(t) + B_2\,e_{i-1,j-1}(t),$$
$$e_{i+1,j}(t^n) = 0, \tag{5.24}$$

$$\frac{de_{i,j+1}(t)}{dt} = A_1\,e_{i,j}(t) + A_2\,e_{i,j+1}(t) + B_1\,e_{i+1,j}(t) + B_2\,e_{i-1,j-1}(t),$$
$$e_{i,j+1}(t^n) = 0, \tag{5.25}$$

$$\frac{de_{i,j}(t)}{dt} = A_1\,e_{i,j}(t) + A_2\,e_{i,j+1}(t) + B_1\,e_{i+1,j}(t) + B_2\,e_{i+1,j+1}(t),$$
$$e_{i,j}(t^n) = 0, \tag{5.26}$$

for $t \in [t^n, t^{n+1}]$, $i, j = 1, 3, 5, \dots$, with $e_{1,1}(0) = 0$ and $e_{1,0}(t) = e_{0,1}(t) = e_{0,0}(t) = c(t)$.

Here we use the notation $\mathbf{X}^4$ for the product space $\Pi_{i=1}^4\mathbf{X}$ enabled with the norm $\|(u_1, u_2, u_3, u_4)^T\| = \max_{i=1,\dots,4}\{\|u_i\|\}$ ($u_i \in \mathbf{X}$, $i = 1, \dots, 4$), and for each component we apply the supremum norm. The elements $\mathcal{E}_i(t)$, $\mathcal{F}_i(t) \in \mathbf{X}^4$ and the linear operator $\mathcal{A} : \mathbf{X}^4 \to \mathbf{X}^4$ are defined as follows:

$$\mathcal{E}_{i,j}(t) = \begin{bmatrix} e_{i,j}(t) \\ e_{i+1,j}(t) \\ e_{i,j+1}(t) \\ e_{i+1,j+1}(t) \end{bmatrix}, \quad \mathcal{A} = \begin{bmatrix} A_1 & 0 & 0 & 0 \\ A_1 & B_1 & 0 & 0 \\ A_1 & B_1 & A_2 & 0 \\ A_1 & B_1 & A_2 & B_2 \end{bmatrix},$$

$$\mathcal{F}_{i,j}(t) = \begin{bmatrix} A_2\,e_{i,j-1}(t) + B_1\,e_{i-1,j}(t) + B_2\,e_{i-1,j-1}(t) \\ A_2\,e_{i,j-1}(t) + B_2\,e_{i-1,j-1}(t) \\ B_2\,e_{i-1,j-1}(t) \\ 0 \end{bmatrix}. \tag{5.27}$$

Using the notations (5.27), the relations (5.23)–(5.26) can be written in the following form:

$$\frac{d\mathcal{E}_{i,j}(t)}{dt} = \mathcal{A}\mathcal{E}_{i,j}(t) + \mathcal{F}_{i,j}(t), \quad t \in [t^n, t^{n+1}],$$
$$\mathcal{E}_{i,j}(t^n) = 0. \tag{5.28}$$

We estimate the right-hand side $\mathcal{F}_{i,j}(t)$ in the following lemma:

LEMMA 5.1
    For $\mathcal{F}_{i,j}(t)$ it holds

$$||\mathcal{F}_{i,j}(t)|| \leq C||e_{i-1,j-1}(t)||.$$

□

PROOF

We have the following norm
$$||\mathcal{F}_{i,j}(t)|| = \max\{||\mathcal{F}_{i,j,1}(t)||, ||\mathcal{F}_{i,j,2}(t)||, ||\mathcal{F}_{i,j,3}(t)||, ||\mathcal{F}_{i,j,4}(t)||\}.$$

Each term can be estimated as

$$||\mathcal{F}_{i,j,1}(t)|| = ||A_2\, e_{i,j-1}(t) + B_1\, e_{i-1,j}(t) + B_2\, e_{i-1,j-1}(t)|| \leq C_1||e_{i-1,j-1}(t)||,$$
$$||\mathcal{F}_{i,j,2}(t)|| = ||A_2\, e_{i,j-1}(t) + B_2\, e_{i-1,j-1}(t)|| \leq C_2||e_{i-1,j-1}(t)||,$$
$$||\mathcal{F}_{i,j,3}(t)|| = ||B_2\, e_{i-1,j-1}(t)|| \leq C_3||e_{i-1,j-1}(t)||.$$

Based on the theorem of Fubini for decouplable operators, see [43], we obtain
$$||e_{\tilde{i},\tilde{j}}(t)|| \leq ||e_{i-1,j-1}||, \text{ for } \tilde{i} = \{i-1, i\} \text{ and } \tilde{j} = \{j-1, j\}.$$
Hence, it holds that
$$||\mathcal{F}_{i,j}(t)|| \leq C||e_{i-1,j-1}(t)||,$$
where C is the maximum value of C_1, C_2 and C_3.

□

Using the variations of constants formula, the solution of the abstract Cauchy problem (5.28) with homogeneous initial conditions can be written as

$$\mathcal{E}_{i,j}(t) = \int_{t^n}^{t} \exp(\mathcal{A}(t-s))\mathcal{F}_{i,j}(s)ds, \quad t \in [t^n, t^{n+1}].$$

(See, e.g., [63].) Hence, using the denotation

$$||\mathcal{F}_{i,j}||_\infty = \sup_{t\in[t^n,t^{n+1}]}||\mathcal{F}_{i,j}(t)||,$$

and taking into account Lemma 5.1, we have

$$||\mathcal{E}_{i,j}(t)|| \leq ||\mathcal{F}_{i,j}||_\infty \int_{t^n}^{t} ||\exp(\mathcal{A}(t-s))||ds$$
$$\leq C\,||e_{i-1,j-1}(t)|| \int_{t^n}^{t} ||\exp(\mathcal{A}(t-s))||ds, \quad t \in [t^n, t^{n+1}].$$

(5.29)

Due to our linearity assumptions for the operators, $\mathcal{A}$ is a generator of the one-parameter C_0 semigroup $(\mathcal{A}(t))_{t\geq 0}$. Because $(\mathcal{A}(t))_{t\geq 0}$ is a semigroup, the *growth estimation*

$$\|\exp(\mathcal{A}t)\| \leq \widetilde{K}\exp(\omega t), \quad t \geq 0, \tag{5.30}$$

holds with some numbers $\widetilde{K} \geq 0$ and $\omega \in \mathbb{R}$, see [63].

- Assume that $(\mathcal{A}(t))_{t\geq 0}$ is a bounded or exponentially stable semigroup that is, (5.30) holds for some $\omega \leq 0$. Then it is apparent that the estimate

$$\|\exp(\mathcal{A}t)\| \leq \widetilde{K}, \quad t \geq 0,$$

  holds, and considering (5.29), we have the relation

$$\|\mathcal{E}_{i,j}(t)\| \leq K\tau_n\|e_{i-1,j-1}(t)\|, \quad t \in [t^n, t^{n+1}]. \tag{5.31}$$

- Assume that $(\mathcal{A}(t))_{t\geq 0}$ has an exponential growth with some $\omega > 0$. Integrating (5.30) yields

$$\int_{t^n}^{t} \|\exp(\mathcal{A}(t-s))\|ds \leq K_\omega(t), \quad t \in [t^n, t^{n+1}], \tag{5.32}$$

  where

$$K_\omega(t) = \frac{\widetilde{K}}{\omega}\left(\exp(\omega(t-t^n)) - 1\right), \quad t \in [t^n, t^{n+1}],$$

  and hence,

$$K_\omega(t) \leq \frac{\widetilde{K}}{\omega}\left(\exp(\omega\tau_n) - 1\right) = \widetilde{K}\tau_n + \mathcal{O}(\tau_n^2), \tag{5.33}$$

  with $\tau_n = t^{n+1} - t^n$.

The estimations (5.29), (5.30), and (5.33) result in

$$\|\mathcal{E}_{i,j}(t)\| \leq K\tau_n\|e_{i-1,j-1}(t)\| + \mathcal{O}(\tau_n^2),$$

where $K = \widetilde{K} \cdot C$ for both cases.

Taking into account the definition of $\mathcal{E}_{i,j}(t)$ and the norm $\|\cdot\|$, we obtain

$$\|e_{i,j}(t)\| \leq K\tau_n\|e_{i-1,j-1}(t)\| + \mathcal{O}(\tau_n^2),$$

which proves our statement.

$\square$

REMARK 5.4 We can generalize our results for n decomposed domains. We obtain the same results for the generalized semigroup $\mathcal{A} : \mathbf{X}^n \to \mathbf{X}^n$. ⬚

REMARK 5.5 We require twice the number of iterations due to the two partitions (i.e., i for time, j for space). Also, for higher-order accuracy more work is needed, for example, $2(2m+1)$ iterations for $\mathcal{O}(\tau^{2m+1})$ accuracy. ⬚

5.2 Schwarz Waveform-Relaxation Methods

5.2.1 Introduction

In the next chapter we discuss the methods for domain decomposition methods.

We motivate our studying on complex models with coupled processes (e.g., transport and reaction equations with nonlinear parameters). The ideas for these models came from the background of the simulation of heat transport in engineering apparatus (e.g., crystal-growth, cf. [95]), or the simulation of chemical reaction and transport (e.g., in bio-remediation or waste disposals, cf. [94]). In the past, many software tools have been developed for one-dimensional and simple physical problems (e.g., one-dimensional transport codes or convection-codes based on ODEs). The future interests will be the coupling of standard software tools to more complex code, for solving multi-dimensional and multiphysical problems.

The mathematical equations are given by

$$\partial_t\, R\, u + \nabla \cdot (\mathbf{v}u - D\nabla u) = f(u)\,, \text{ in } \Omega \times (0, T), \tag{5.34}$$

$$u(x, 0) = u_0\,, \text{ (Initial Condition)}, \tag{5.35}$$

$$u(x, t) = u_1\,, \text{ on } \partial\Omega \times (0, T)\,, \text{ (Dirichlet-Boundary Condition)}, \tag{5.36}$$

$$f(u) = u^p\,, \text{ chemical reaction and } p > 0, \tag{5.37}$$

$$f(u) = \frac{u}{1 - u}\,, \text{ bio remediation}, \tag{5.38}$$

$$f(u) = f(u_0)\,, \text{ heat induction}. \tag{5.39}$$

The unknown $u = u(x, t)$ is considered in $\Omega \times (0, T) \subset \mathbb{R}^d \times \mathbb{R}$, the space dimension is given by d. The parameter $R \in \mathbb{R}^+$ is a constant and named as retardation factor. The parameters $u_0, u_1 \in \mathbb{R}^+$ are constants and used as initial- and boundary-parameter, respectively. The parameter $f(u)$ is a nonlinear function, for example, bio remediation or chemical reaction. D is the thermal conductivity tensor or Scheidegger diffusion-dispersion tensor, and **v** is the velocity.

Our aim is to present a new method based on a mixed discretization method with fractional splitting and domain decomposition methods for an effective solving of strong coupled parabolic differential equations.

In the next section we discuss the fractional splitting methods for solving our equations.

Now we introduce the domain-decomposition methods as the next idea for splitting methods to decompose complex domains and solve them effectively in an adaptive method.

5.3 Overlapping Schwarz Waveform Relaxation for the Solution of Convection-Diffusion-Reaction Equation

The first known method for solving a partial differential equation over overlapped domains is the Schwarz method due to [180] in 1869. The method has regained its popularity after the development of the computational numerical algorithms and the computer architecture, especially the parallel processing computations.

Further techniques have been developed for the general cases when the domains are overlapped and nonoverlapped. For each class of methods there are some interesting features, and both share the same concepts on how to define the interface boundary conditions over the overlapped or along the nonoverlapped subdomains. The general solution methods over the whole subdomains together with the interface boundary conditions estimations are either iterative or noniterative methods.

For the nonoverlapping subdomains, the interfaces are predicted using explicit scheme and the problem is solved over each subdomain independently. This type of method is of the noniterative type, but it has drawbacks regarding the stability condition for the interface prediction by the explicit method and the solution by the implicit scheme or any other unconditional stable finite difference scheme [57].

For the overlapping subdomains, the determination of the interface boundary condition is defined using predictor corrector type of method. The predictor will provide an estimation of the boundary condition while the correction is performed from the updated solution over the subdomains. This class of algorithms is of an iterative type with the advantage that the stability of the solution by any unconditional difference approximation will not be affected by the predicted interface values.

In this work we will consider the overlapping type of domain decomposition method for solving the studied models of constant coefficients, decoupled and coupled systems solved using the first-order operator-splitting algorithm with backward Euler difference scheme. The most recent method in this field is the overlapping Schwarz waveform relaxation scheme due to [77] and [107].

Overlapping Schwarz waveform relaxation is the name for a combination of two standard algorithms, Schwarz alternating method and waveform-relaxation algorithm to solve evolution problems in parallel. The method is defined by partitioning the spatial domain into overlapping subdomains, as in the classical Schwarz method. However, on subdomains, time-dependent problems are solved in the iteration and thus the algorithm is also of waveform-relaxation type. Furthermore, the problem is solved using the operator splitting of first-order over each subdomain. The overlapping Schwarz waveform relaxation is introduced in [107] and independently in [77] for the solver method of evolution problems in a parallel environment with slow communication links. The idea is to solve over several time-steps before communicating information to the neighboring subdomains and updating the calculated interface boundary conditions for the overlapped domains.

These algorithms contrast with the classical approach in domain decomposition for evolution problems, where time is first discretized uniformly using an implicit discretization and then at each time-step a problem in space only is solved using domain decomposition, see for example [35], [36], and [157]. Furthermore, in this work the operator-splitting method will be considered by using Crank-Nicolson (CN) or an implicit Euler method for the time discretization. The main advantage in considering the overlapping Schwarz wave form relaxation method is the flexibility that we could solve over each subdomain with a different time-step and different spatial steps in the whole time interval. In this section we will consider the Schwarz waveform relaxation to solve scalar, and systems of convection-diffusion-reaction equation. For the systems of convection-diffusion-reaction equation, we study the decoupled case, i.d. m scalar equations and the coupled case, i.d. m equations coupled by the reaction terms.

5.3.1 Overlapping Schwarz Waveform Relaxation for the Scalar Convection-Diffusion-Reaction Equation

We consider the convection-diffusion-reaction equation, given by

$$u_t = Du_{xx} - vu_x - \lambda u , \qquad (5.40)$$

defined on the domain $\Omega = [0, L]$ for $T = [t_0, T_f]$, with the following initial and boundary conditions:

$$u(0,t) = f_1(t), \quad u(L,t) = f_2(t), \quad u(x,t_0) = u_0.$$

To solve the model problem using the overlapping Schwarz waveform relaxation method, we subdivide the domain Ω in two overlapping subdomains $\Omega_1 = [0, L_2]$ and $\Omega_2 = [L_1, L]$, where $L_1 < L_2$ and $\Omega_1 \bigcap \Omega_2 = [L_1, L_2]$ is the overlapping region for Ω_1 and Ω_2.

To start the waveform relaxation algorithm, we first consider the solution of the model problem (5.40) over Ω_1 and Ω_2 as follows:

$$
\begin{aligned}
v_t &= Dv_{xx} - vv_x - \lambda v \text{ over } \Omega_1 , \quad t \in [T_0, T_f], \\
v(0,t) &= f_1(t) , \quad t \in [T_0, T_f], \\
v(L_2,t) &= w(L_2,t) , \quad t \in [T_0, T_f], \\
v(x,t_0) &= u_0 \quad x \in \Omega_2,
\end{aligned}
\tag{5.41}
$$

$$
\begin{aligned}
w_t &= Dw_{xx} - vw_x - \lambda w \text{ over } \Omega_2 , \quad t \in [T_0, T_f], \\
w(L_1,t) &= v(L_1,t) , \quad t \in [T_0, T_f], \\
w(L,t) &= f_2(t) , \quad t \in [T_0, T_f], \\
w(x,t_0) &= u_0 \quad x \in \Omega_2,
\end{aligned}
\tag{5.42}
$$

where $v(x,t) = u(x,t)|_{\Omega_1}$ and $w(x,t) = u(x,t)|_{\Omega_2}$.

Then the Schwarz waveform relaxation is given by

$$
\begin{aligned}
v_t^{k+1} &= Dv_{xx}^{k+1} - vv_x^{k+1} - \lambda v^{k+1} \text{ over } \Omega_1 , \quad t \in [T_0, T_f], \\
v^{k+1}(0,t) &= f_1(t) , \quad t \in [T_0, T_f], \\
v^{k+1}(L_2,t) &= w^k(L_2,t) , \quad t \in [T_0, T_f], \\
v^{k+1}(x,t_0) &= u_0 \quad x \in \Omega_1,
\end{aligned}
\tag{5.43}
$$

$$
\begin{aligned}
w_t^{k+1} &= Dw_{xx}^{k+1} - vw_x^{k+1} - \lambda w^{k+1} \text{ over } \Omega_2 , \quad t \in [T_0, T_f], \\
w^{k+1}(L_1,t) &= v^k(L_1,t) , \quad t \in [T_0, T_f], \\
w^{k+1}(L,t) &= f_2(t) , \quad t \in [T_0, T_f], \\
w^{k+1}(x,t_0) &= u_0 \quad x \in \Omega_2.
\end{aligned}
\tag{5.44}
$$

We are interested in estimating the decay of the error of the solution over the overlapping subdomains by the overlapping Schwarz waveform relaxation method.

Let us assume $e(x,t) = u(x,t) - v(x,t)$, and $d(x,t) = u(x,t) - w(x,t)$ is the error of (5.43) and (5.44) over Ω_1 and Ω_2, respectively. The corresponding differential equations satisfying $e(x,t)$ and $d(x,t)$ are given by

$$
\begin{aligned}
e_t^{k+1} &= De_{xx}^{k+1} - ve_x^{k+1} - \lambda e^{k+1} \text{ over } \Omega_1 , \quad t \in [T_0, T_f], \\
e^{k+1}(0,t) &= 0 , \quad t \in [T_0, T_f], \\
e^{k+1}(L_2,t) &= d^k(L_2,t) , \quad t \in [T_0, T_f], \\
e^{k+1}(x,t_0) &= 0 \quad x \in \Omega_1,
\end{aligned}
\tag{5.45}
$$

$$
\begin{aligned}
d_t^{k+1} &= Dd_{xx}^{k+1} - vd_x^{k+1} - \lambda d^{k+1} \text{ over } \Omega_2 , \quad t \in [T_0, T_f], \\
d^{k+1}(L_1,t) &= e^k(L_1,t) , \quad t \in [T_0, T_f], \\
d^{k+1}(L,t) &= 0 , \quad t \in [T_0, T_f], \\
d^{k+1}(x,t_0) &= 0 , \quad x \in \Omega_2.
\end{aligned}
\tag{5.46}
$$

We define for bounded functions $h(x,t) : \Omega \times [0,T] \to \mathbb{R}$ the norm

$$
||h(.,.)||_\infty := \sup_{x \in \Omega, t \in [t_0, T_f]} |h(x,t)|.
$$

The convergence and error bound of e^{k+1} and d^{k+1} are presented in Theorem 5.2.

THEOREM 5.2
Let e^{k+1} and d^{k+1} be the error from the solution of the subproblems (5.41) and (5.42) by Schwarz waveform relaxation over Ω_1 and Ω_2, respectively, then

$$||e^{k+2}(L_1,t)||_\infty \leq \gamma||e^k(L_1,t)||_\infty,$$

and

$$||d^{k+2}(L_1,t)||_\infty \leq \gamma||d^k(L_1,t)||_\infty,$$

where

$$\gamma = \frac{\sinh(\beta L_1)}{\sinh(\beta L_2)}\frac{\sinh(\beta(L_2-L))}{\sinh(\beta(L_1-L))}.$$

PROOF

For the error e^{k+1} and d^{k+1}, consider the following differential equations defined by $\hat{e}^{k+1}$ and $\hat{d}^{k+1}$ given by

$$
\begin{aligned}
\hat{e}^{k+1}_t &= D\hat{e}^{k+1}_{xx} - v\hat{e}^{k+1}_x - \lambda\hat{e}^{k+1} \text{ over } \Omega_1, \quad t \in [T_0, T_f],\\
\hat{e}^{k+1}(0,t) &= 0, \quad t \in [T_0, T_f],\\
\hat{e}^{k+1}(L_2,t) &= ||d^k(L_2,t)||_\infty, \quad t \in [T_0, T_f],\\
\hat{e}^{k+1}(x,t_0) &= e^{(x-L_2)\alpha}\frac{\sinh(\beta x)}{\sinh(\beta L_2)}||d^k(L_2,t)||_\infty, \quad x \in \Omega_1,
\end{aligned}
\tag{5.47}
$$

and

$$
\begin{aligned}
\hat{d}^{k+1}_t &= D\hat{d}^{k+1}_{xx} - v\hat{d}^{k+1}_x - \lambda\hat{d}^{k+1} \text{ over } \Omega_2, \quad t \in [T_0, T_f],\\
\hat{d}^{k+1}(L_1,t) &= ||e^k(L_1,t)||_\infty, \quad t \in [T_0, T_f],\\
\hat{d}^{k+1}(L,t) &= 0, \quad t \in [T_0, T_f],\\
\hat{d}^{k+1}(x,t_0) &= e^{(x-L_1)\alpha}\frac{\sinh\beta(x-L)}{\sinh\beta(L_1-L)}||e^k(L_1,t)||_\infty, \quad x \in \Omega_2,
\end{aligned}
\tag{5.48}
$$

where $\alpha = \frac{v}{2D}$ and $\beta = \frac{\sqrt{v^2+4D\lambda}}{2D}$.

The solution to (5.47) and (5.48) is the steady state solution given by

$$\hat{e}^{k+1}(x) = e^{(x-L_2)\alpha}\frac{\sinh(\beta x)}{\sinh(\beta L_2)}||d^k(L_2,t)||_\infty,$$

and

$$\hat{d}^{k+1}(x) = e^{(x-L_1)\alpha}\frac{\sinh\beta(x-L)}{\sinh\beta(L_1-L)}||e^k(L_1,t)||_\infty,$$

respectively.

Hence, define $E(x,t) = \hat{e}^{k+1} - e^{k+1}$, and therefore,

$$
\begin{aligned}
E_t - DE_{xx} + vE_x + \lambda E &\geq 0 \ , \ \text{over } \Omega_1 \ , \quad t \in [T_0, T_f], \\
E(0,t) &= 0 \ , \quad t \in [T_0, T_f], \\
E(L_2,t) &\geq 0 \ , \quad t \in [T_0, T_f], \\
E(x,t_0) &\geq 0 \ , \quad x \in \Omega_1,
\end{aligned}
\tag{5.49}
$$

and it satisfies the positivity lemma by Pao (or the maximum principle theorem), see [164]; therefore,

$$E(x,t) \geq 0$$

that is,

$$|e^{k+1}| \leq \hat{e}^{k+1} \ ,$$

for all (x,t), and similarly we conclude that

$$|d^{k+1}| \leq \hat{d}^{k+1} \ ,$$

for all (x,t).
  Then

$$|e^{k+1}(x,t)| \leq e^{(x-L_2)\alpha}\frac{\sinh(\beta x)}{\sinh(\beta L_2)}||d^k(L_2,t)||_\infty \ , \tag{5.50}$$

and

$$|d^{k+1}(x,t)| \leq e^{(x-L_1)\alpha}\frac{\sinh\beta(x-L_1)}{\sinh\beta(L_1-L)}||e^k(L_1,t)||_\infty \ . \tag{5.51}$$

Evaluate $d^k(x,t)$ at L_2:

$$|d^k(L_2,t)| \leq \frac{\sinh\beta(L_2-L)}{\sinh\beta(L_1-L)}||e^{k-1}(L_1,t)||_\infty \ , \tag{5.52}$$

and substitute in (5.50) implies that

$$|e^{k+1}(x,t)| \leq e^{(x-L_2)\alpha}\frac{\sinh(\beta x)}{\sinh(\beta L_2)}e^{(L_2-L_1)\alpha}\frac{\sinh\beta(L_2-L)}{\sinh\beta(L_1-L)}||e^{k-1}(L_1,t)||_\infty \ ,$$

therefore

$$|e^{k+1}(L_1,t)| \leq e^{(L_1-L_2)\alpha}\frac{\sinh(\beta L_1)}{\sinh(\beta L_2)}e^{(L_2-L_1)\alpha}\frac{\sinh\beta(L_2-L)}{\sinh\beta(L_1-L)}||e^{k-1}(L_1,t)||_\infty \ ,$$

that is,

$$|e^{k+2}(L_1,t)| \leq \frac{\sinh(\beta L_1)}{\sinh(\beta L_2)}\frac{\sinh\beta(L_2-L)}{\sinh\beta(L_1-L)}||e^k(L_1,t)||_\infty \ .$$

Similarly for $d^{k+1}(x,t)$ we conclude that

$$|d^{k+2}(L_1,t)| \leq \frac{\sinh(\beta L_1)}{\sinh(\beta L_2)} \frac{\sinh \beta(L_2 - L)}{\sinh \beta(L_1 - L)} ||d^k(L_1,t)||_\infty .$$

☐

Theorem 5.2 shows that the convergence of of the overlapping Schwarz method depends on $\gamma = \frac{\sinh(\beta L_1)}{\sinh(\beta L_2)} \frac{\sinh \beta(L_2-L)}{\sinh \beta(L_1-L)}$. Due to the characteristic of the sinh function, we will have sharp decay of the error for any $L_1 < L_2$, and also for a large amount of overlapping.

5.3.2 Overlapping Schwarz Waveform Relaxation for Decoupled System of Convection-Diffusion-Reaction Equation

In the following part of this section, we are going to present, overlapping Schwarz waveform-relaxation method defined for a system of convection-diffusion-reaction equation, such that for a system defined by $u_i(x,t)$ for $i = 1, \ldots, I$ it is given by

$$
\begin{aligned}
R_i u_{i,t} &= D_i u_{i,xx} - v_i u_{i,x} - \lambda_i u_i \text{ over } \Omega, \ t \in [T_0, T_f], \\
u_i(t_0, t) &= f_{1,i}(t), \ t \in [T_0, T_f], \\
u_i(L, t) &= f_{2,i}(L, t), \ t \in [T_0, T_f], \\
u_i(x, t_0) &= u_0, \ x \in \Omega.
\end{aligned}
\tag{5.53}
$$

The considered system (5.53) is already defined over the spatial domain $\Omega = \{0 < x < L\}$ and the overlapping Schwarz over relaxation method is constructed over the overlapping subdomains $\Omega_1 = \{0 < x < L_2\}$ and $\Omega_2 = \{L_1 < x < L\}$ $L_1 < L_2$ with an overlapping size $(L2 - L1)$. In this work we are going to consider two types of systems of convection-diffusion-reaction equation, the decoupled and coupled systems.

To construct the waveform-relaxation algorithm for (5.53), we will consider the case where $I = 2$, and we first consider the solution of (5.53) over Ω_1 and Ω_2 as follows:

$$
\begin{aligned}
R_i v_{i,t} &= D_i v_{i,xx} - v_i v_{i,x} - \lambda_i v_i \text{ over } \Omega_i, \ t \in [T_0, T_f], \\
v_i(t_0, t) &= f_{1,i}(t), \ t \in [T_0, T_f], \\
v_i(L, t) &= f_{2,i}(L, t), \ t \in [T_0, T_f], \\
v_i(x, t_0) &= u_0, \ x \in \Omega_i,
\end{aligned}
\tag{5.54}
$$

$$
\begin{aligned}
R_i w_{i,t} &= D_i w_{i,xx} - v_i w_{i,x} - \lambda_i w_i \text{ over } \Omega_i, \ t \in [T_0, T_f], \\
w_i(t_0, t) &= f_{1,i}(t), \ t \in [T_0, T_f], \\
w_i(L, t) &= f_{2,i}(L, t), \ t \in [T_0, T_f], \\
w_i(x, t_0) &= u_0, \ x \in \Omega_i,
\end{aligned}
\tag{5.55}
$$

where $v_i(x,t) = u_i(x,t)|\Omega_1$ and $w_i(x,t) = u_i(x,t)|\Omega_2$.

Then the overlapping Schwarz waveform-relaxation method for the decoupled system over the two overlapped subdomains, Ω_1 and Ω_2, is given by

$$
\begin{aligned}
R_1 v_{1,t}^{k+1} &= D_1 v_{1,xx}^{k+1} - \mathbf{v}_1 v_{1,x}^{k+1} - \lambda_1 v_1^{k+1} \text{ over } \Omega_1, \quad t \in [T_0, T_f], \\
v_1^{k+1}(t_0, t) &= f_{1,1}(t), \quad t \in [T_0, T_f], \\
v_1^{k+1}(L_2, t) &= w_1^k(L_2, t), \quad t \in [T_0, T_f], \\
v_1^{k+1}(x, t_0) &= u_0, \quad x \in \Omega_1,
\end{aligned}
\tag{5.56}
$$

$$
\begin{aligned}
R_1 w_{1,t}^{k+1} &= D_1 w_{1,xx}^{k+1} - \mathbf{v}_1 w_{1,x}^{k+1} - \lambda_1 w_1^{k+1} \text{ over } \Omega_2, \quad t \in [T_0, T_f], \\
w_1^{k+1}(L_1, t) &= v_1^k(L_1, t), \quad t \in [T_0, T_f], \\
w_1^{k+1}(L, t) &= f_{1,2}(t), \quad t \in [T_0, T_f], \\
w_1^{k+1}(x, t_0) &= u_0, \quad x \in \Omega_2,
\end{aligned}
\tag{5.57}
$$

for the system defined by u_1 for $i = 1$, and for the system defined by u_2, $i = 2$, it is given by

$$
\begin{aligned}
R_2 v_{2,t}^{k+1} &= D_2 v_{2,xx}^{k+1} - \mathbf{v}_2 v_{2,x}^{k+1} - \lambda_2 v_2^{k+1} \text{ over } \Omega_1, \quad t \in [T_0, T_f], \\
v_2^{k+1}(t_0, t) &= f_{2,1}(t), \quad t \in [T_0, T_f], \\
v_2^{k+1}(L_2, t) &= w_2^k(L_2, t), \quad t \in [T_0, T_f], \\
v_2^{k+1}(x, t_0) &= u_0, \quad x \in \Omega_1,
\end{aligned}
\tag{5.58}
$$

$$
\begin{aligned}
R_2 w_{2,t}^{k+1} &= D_2 w_{2,xx}^{k+1} - \mathbf{v}_2 w_{2,x}^{k+1} - \lambda_2 w_2^{k+1} \text{ over } \Omega_2, \quad t \in [T_0, T_f], \\
w_2^{k+1}(L_1, t) &= v_2^k(L_1, t), \quad t \in [T_0, T_f], \\
w_2^{k+1}(L, t) &= f_{2,2}(t), \quad t \in [T_0, T_f], \\
w_2^{k+1}(x, t_0) &= u_0 \quad x \in [T_0, T_f],
\end{aligned}
\tag{5.59}
$$

where k represents the iteration index.

Define $e_i^{k+1} = u - v_i^{k+1}$ and $d_i^{k+1} = u - w_i^{k+1}$ to be the errors from the solution given by (5.56)–(5.57) and (5.58)–(5.59) over Ω_1 and Ω_2, respectively, and for $i = 1, 2$.

The corresponding differential equations satisfying e_i^{k+1} and d_i^{k+1} over Ω_1 and Ω_2 for $i = 1, 2$, respectively, are given by

$$
\begin{aligned}
R_1 e_{1,t}^{k+1} &= D_1 e_{1,xx}^{k+1} - \mathbf{v}_1 e_{1,x}^{k+1} - \lambda_1 e_1^{k+1} \text{ over } \Omega_1, \quad t \in [T_0, T_f], \\
e_1^{k+1}(t_0, t) &= f_{1,1}(t), \quad t \in [T_0, T_f], \\
e_1^{k+1}(L_1, t) &= d_1^k(L_1, t), \quad t \in [T_0, T_f], \\
e_1^{k+1}(x, t_0) &= u_0, \quad x \in \Omega_1,
\end{aligned}
\tag{5.60}
$$

$$
\begin{aligned}
R_1 d_{1,t}^{k+1} &= D_1 d_{1,xx}^{k+1} - \mathbf{v}_1 d_{1,x}^{k+1} - \lambda_1 d_1^{k+1} \text{ over } \Omega_2, \quad t \in [T_0, T_f], \\
d_1^{k+1}(L_2, t) &= e_1^k(L_2, t), \quad t \in [T_0, T_f], \\
d_1^{k+1}(L, t) &= f_{1,2}(t), \quad t \in [T_0, T_f], \\
d_1^{k+1}(x, t_0) &= u_0 \quad x \in [T_0, T_f],
\end{aligned}
\tag{5.61}
$$

for $i = 1$, and for $i = 2$,

$$
\begin{aligned}
R_2 e_{2,t}^{k+1} &= D_2 e_{2,xx}^{k+1} - v_2 e_{2,x}^{k+1} - \lambda_2 e_2^{k+1} \text{ over } \Omega_1, \ t \in [T_0, T_f], \\
e_2^{k+1}(t_0, t) &= f_{2,1}(t), \ t \in [T_0, T_f], \\
e_2^{k+1}(L_1, t) &= d_2^k(L_1, t), \ t \in [T_0, T_f], \\
e_2^{k+1}(x, t_0) &= u_0, \ x \in \Omega_1,
\end{aligned}
\tag{5.62}
$$

$$
\begin{aligned}
R_2 d_{2,t}^{k+1} &= D_2 d_{2,xx}^{k+1} - v_2 d_{2,x}^{k+1} - \lambda_2 d_2^{k+1} \text{ over } \Omega_2, \ t \in [T_0, T_f], \\
d_2^{k+1}(L_1, t) &= e_2^k(L_1, t), \ t \in [T_0, T_f], \\
d_2^{k+1}(L, t) &= f_{2,2}(t), \ t \in [T_0, T_f], \\
d_2^{k+1}(x, t_0) &= u_0 \ x \in [T_0, T_f].
\end{aligned}
\tag{5.63}
$$

The convergence and the error bound for the solution of (5.56)–(5.57) and (5.58)–(5.59) are given in Theorem 5.3.

THEOREM 5.3

Let e_i^{k+1} and d_i^{k+1} $(i = 1, 2)$ be the error of the subproblems defined by the differential equations (5.60)–(5.61) and (5.62)–(5.63) over Ω_1 and Ω_1, respectively, then

$$
||e_i^{k+2}(L_1, t)||_\infty \le \gamma_i ||e_i^k(L_1, t)||_\infty,
\tag{5.64}
$$

and

$$
||d_i^{k+2}(L_1, t)||_\infty \le \gamma_i ||d_i^k(L_1, t)||_\infty,
\tag{5.65}
$$

where

$$
\gamma_i = \frac{\sinh(\beta_i L_1)}{\sinh(\beta_i L_2)} \frac{\sinh(\beta_i(L_2 - L))}{\sinh(\beta_i(L_1 - L))},
\tag{5.66}
$$

for $i = 1, 2$.

PROOF

The proof will follow by utilization the proof given by Theorem 5.2.

Let e_i and d_i be the error of the approximated solutions u_i over Ω_1 and Ω_2, for $i = 1, 2$, respectively.

Following the proof presented for Theorem 5.2 for each of the error differential equations (5.60)–(5.61) and (5.62)–(5.63), then we will conclude the following relation:

$$
||e_i^{k+2}(L_1, t)||_\infty \le \gamma_i ||e_i^k(L_1, t)||_\infty,
\tag{5.67}
$$

and

$$
||d_i^{k+2}(L_1, t)||_\infty \le \gamma_i ||d_i^k(L_1, t)||_\infty,
\tag{5.68}
$$

where

$$\gamma_i = \frac{\sinh(\beta_i L_1)}{\sinh(\beta_i L_2)} \frac{\sinh(\beta_i(L_2 - L))}{\sinh(\beta_i(L_1 - L))},$$

(5.69)

and

$$\beta_i = \frac{\sqrt{v_i^2 + 4D_i\lambda_i}}{2D_i}$$

for $i = 1, 2$.

☐

5.3.3 Overlapping Schwarz Waveform Relaxation for Coupled System of Convection-Diffusion-Reaction Equation

In the following part we are going to present the convergence and the error bound of the overlapping Schwarz waveform relaxation for the solution of the coupled system of convection-diffusion-reaction defined by two functions u_1 and u_2. The coupling criterion in this case of study is imposed within the source term of the second solution component. The considered system with the solution u_1 and u_2 is given by

$$
\begin{aligned}
R_1 u_{1,t} &= D_1 u_{1,xx} - v_1 u_{1,x} - \lambda_1 u_1 \text{ over } \Omega = \{0 < x < L\}, \quad t \in [T_0, T_f], \\
u_1(t_0, t) &= f_{1,1}(t), \quad t \in [T_0, T_f], \\
u_1(L, t) &= f_{1,2}(t), \quad t \in [T_0, T_f], \\
u_1(x, t_0) &= u_0,
\end{aligned}
$$

(5.70)

for u_1, and for u_2 is given by

$$
\begin{aligned}
R_2 u_{2,t} &= D_2 u_{2,xx} - v_2 u_{2,x} - \lambda_2 u_2 + \lambda_1 u_1 \text{ over } \Omega, \quad t \in [T_0, T_f] \\
u_2(t_0, t) &= f_{2,1}(t), \quad t \in [T_0, T_f], \\
u_2(L, t) &= f_{2,2}(t), \quad t \in [T_0, T_f], \\
u_2^{k+1}(x, t_0) &= u_0.
\end{aligned}
$$

(5.71)

In (5.71) the coupling appeared in the source term and is defined by the parameter λ_1 with the first component u_1. The strength or the *bound* of the coupling and the contribution are related to the value of the scalar defined by λ_1. The coupled case (5.71) is reduced to the decoupled case (5.53), by assuming $\lambda_1 = 0$.

The overlapping Schwarz waveform relaxation for (5.70) over Ω_1 and Ω_2 is given by

$$
\begin{aligned}
R_1 v_{1,t}^{k+1} &= D_1 v_{1,xx}^{k+1} - v_1 v_{1,x}^{k+1} - \lambda_1 v_1^{k+1} \text{ over } \Omega_1, \quad t \in [T_0, T_f], \\
v_1^{k+1}(t_0, t) &= f_{1,1}(t), \quad t \in [T_0, T_f], \\
v_1^{k+1}(L_2, t) &= w_1^k(L_2, t), \quad t \in [T_0, T_f], \\
v_1^{k+1}(x, t_0) &= u_0, x \in \Omega_1,
\end{aligned}
$$

(5.72)

$$\begin{aligned}
R_1 w_{1,t}^{k+1} &= D_1 w_{1,xx}^{k+1} - v_1 w_{1,x}^{k+1} - \lambda_1 w_1^{k+1} \text{ over } \Omega_2, \quad t \in [T_0, T_f], \\
w_1^{k+1}(L_1, t) &= v_1^k(L_1, t), \quad t \in [T_0, T_f], \\
w_1^{k+1}(L, t) &= f_{1,2}(t), \quad t \in [T_0, T_f], \\
w_1^{k+1}(x, t_0) &= u_0, x \in \Omega_2,
\end{aligned} \tag{5.73}$$

and for the system defined by (5.71) we denote the Schwarz waveform relaxation as

$$\begin{aligned}
R_2 v_{2,t}^{k+1} &= D_2 v_{2,xx}^{k+1} - v_2 v_{2,x}^{k+1} \\
&\quad -\lambda_2 v_2^{k+1} + \lambda_1 v_1^{k+1} \text{ over } \Omega_1, \quad t \in [T_0, T_f], \\
v_2^{k+1}(0, t) &= f_{2,1}(t), \quad t \in [T_0, T_f], \\
v_2^{k+1}(L_2, t) &= w_2^k(L_2, t), \quad t \in [T_0, T_f], \\
v_2^{k+1}(x, t_0) &= u_0, \quad x \in \Omega_1,
\end{aligned} \tag{5.74}$$

$$\begin{aligned}
R_2 w_{2,t}^{k+1} &= D_2 w_{2,xx}^{k+1} - v_2 w_{2,x}^{k+1} \\
&\quad -\lambda_2 w_2^{k+1} + \lambda_1 w_1^{k+1} \text{ over } \Omega_2, \quad t \in [T_0, T_f], \\
w_2^{k+1}(L_1, t) &= v_2^k(L_1, t), \quad t \in [T_0, T_f], \\
w_1^{k+1}(L, t) &= f_{2,2}(t), \quad t \in [T_0, T_f], \\
w_1^{k+1}(x, t_0) &= u_0 \quad x \in \Omega_2.
\end{aligned} \tag{5.75}$$

The convergence and the error bound for the solution of (5.72)–(5.73) and (5.74)–(5.75) are given by the following theorem.

THEOREM 5.4

Let e_i^{k+1} and d_i^{k+1} $(i = 1, 2)$ be the errors from the solution of the subproblems (5.72)–(5.73) and (5.74)–(5.75) by Schwarz waveform relaxation over Ω_1 and Ω_2, respectively. Then the error bounds of (5.72)-(5.73) defined by e_1 and d_1 over Ω_1 and Ω_2 are given by

$$||e_1^{k+2}(L_1, t)||_\infty \leq \gamma_1 ||e_1^k(L_1, t)||_\infty, \tag{5.76}$$

and

$$||d_1^{k+2}(L_1, t)||_\infty \leq \gamma_1 ||d_1^k(L_1, t)||_\infty, \tag{5.77}$$

respectively, and the error bound of (5.74- 5.75) defined by e_2 and d_2 over Ω_1 and Ω_2 are given by

$$||e_2^{k+2}(L_1, t)||_\infty \leq ||e_2^k(L_1, t)||_\infty \gamma_2 + \gamma_2 \frac{\lambda_1}{\lambda_2} \Psi \left[1 + e^{\alpha_2(L_1 - L)} e^{\beta_2(L - L_1)} \right]$$

$$+ \frac{\lambda_1}{\lambda_2} \Psi \left[e^{\alpha_2(L_1 - L_2)} \frac{\sinh \beta_2 L_1}{\sinh \beta_2 L_2} - e^{\alpha_2(L_1 - L)} e^{\beta_2(L - L_2)} \frac{\sinh \beta_2 L_1}{\sinh \beta_2 L_2} \right] +$$

$$\frac{\lambda_1}{\lambda_2} \Psi \left[e^{\alpha_2 L_1} \frac{\sinh \beta_2(L_1 - L_2)}{\sinh \beta_2 L_2} - e^{\alpha_2(L_1 - L_2)} \frac{\sinh \beta_2 L_1}{\sinh \beta_2 L_2} + 1 \right], \tag{5.78}$$

and

$$||d_2^{k+2}(L_2,t)||_\infty \leq ||d_2^k(L_2,t)||_\infty \gamma_2 + \gamma_2 \frac{\lambda_1}{\lambda_2}\Psi \left[1 + e^{\alpha_2(L_1-L)}e^{\beta_2(L-L_1)}\right]$$

$$+\frac{\lambda_1}{\lambda_2}\Psi\left[e^{\alpha_2(L_1-L_2)}\frac{\sinh \beta_2 L_1}{\sinh \beta_2 L_2} - e^{\alpha_2(L_1-L)}e^{\beta_2(L-L_2)}\frac{\sinh \beta_2 L_1}{\sinh \beta_2 L_2}\right] +$$

$$\frac{\lambda_1}{\lambda_2}\Psi\left[e^{\alpha L_1}\frac{\sinh \beta_2(L_1-L_2)}{\sinh \beta_2 L_2} - e^{\alpha_2(L_1-L_2)}\frac{\sinh \beta_2 L_1}{\sinh \beta_2 L_2} + 1\right],$$

$$(5.79)$$

respectively, where

$$\gamma_i = \frac{\sinh \beta_i L_1}{\sinh \beta_i L_2}\frac{\sinh \beta_i(L_2 - L)}{\sinh \beta_i(L_1 - L)},$$

with $\alpha_i = \frac{v_i}{2D_i}$, $\beta_i = \frac{\sqrt{v_i^2 + 4D_i\lambda_i}}{2D_i}$, *for* $i = 1,2$, *and* $\Psi = \max_\Omega \{e_1, e_2\}$.

PROOF

For the proof of (5.76) and (5.77) they follow from the proof given by Theorem 5.3 for the decoupling case of system.

Let $e_2^{k+1}(x,t) := u_2(x,t) - v_2^{k+1}(x,t)$ and $d_2^{k+1}(x,t) := u_2(x,t) - w_2^{k+1}(x,t)$ be the error of (5.74) and (5.75) over Ω_1 and Ω_2, respectively. Then the corresponding differential equations defined by $e_2(x,t)$ and $d_2(x,t)$ are given by

$$\begin{aligned}
R_2 e_{2,t}^{k+1} &= D_2 e_{2,xx}^{k+1} - v_2 e_{2,x}^{k+1} \\
&\quad -\lambda_2 e_2^{k+1} + \lambda_1 e_1^{k+1} \text{ over } \Omega_1, \ t \in [T_0, T_f], \\
e_2^{k+1}(t_0, t) &= 0, \ t \in [T_0, T_f], \\
e_2^{k+1}(L_2, t) &= d_2^k(L_2, t), \ t \in [T_0, T_f], \\
e_2^{k+1}(x, t_0) &= 0, \ x \in \Omega_2,
\end{aligned} \tag{5.80}$$

$$\begin{aligned}
R_2 d_{2,t}^{k+1} &= D_2 d_{2,xx}^{k+1} - v_2 d_{2,x}^{k+1} \\
&\quad -\lambda_2 d_2^{k+1} + \lambda_1 d_1^{k+1} \text{ over } \Omega_2, \ t \in [T_0, T_f], \\
d_2^{k+1}(L_1, t) &= e_2^k(L_1, t), \ t \in [T_0, T_f], \\
d_1^{k+1}(L, t) &= 0, \ t \in [T_0, T_f], \\
d_1^{k+1}(x, t_0) &= 0, \ x \in \Omega_2.
\end{aligned} \tag{5.81}$$

Furthermore, we consider the following differential equations defined by $\hat{e}^{k+1}$ and $\hat{d}^{k+1}$

$$\begin{aligned}
R_2 \hat{e}_{2,t}^{k+1} &= D_2 \hat{e}_{2,xx}^{k+1} - v_2 \hat{e}_{2,x}^{k+1} \\
&\quad -\lambda_2 \hat{e}_2^{k+1} + \lambda_1 \Psi \text{ over } \Omega_1, \ t \in [T_0, T_f], \\
\hat{e}_2^{k+1}(t_0, t) &= 0, \ t \in [T_0, T_f], \\
\hat{e}_2^{k+1}(L_2, t) &= ||d_2^k(L_2, t)||_\infty, \ t \in [T_0, T_f], \\
\hat{e}_2^{k+1}(x, t_0) &= \mathcal{A}(x), \ x \in \Omega_1,
\end{aligned} \tag{5.82}$$

where

$$A(x) = ||d_2^k(L_2, t)||_\infty e^{\alpha_2(x-L_2)} \frac{\sinh(\beta_2 x)}{\sinh(\beta_2 L)}$$

$$+ \frac{\lambda_1}{\lambda_2} \Psi \left[e^{\alpha_2 x} \frac{\sinh(\beta_2(x-L_2))}{\sinh(\beta_2 L_2)} - e^{\alpha_2(x-L_2)} \frac{\sinh \beta_2 x}{\sinh \beta_2 L_2} + 1 \right], \quad (5.83)$$

and

$$\begin{aligned}
R_2 \hat{d}_{2,t}^{k+1} &= D_2 \hat{d}_{2,xx}^{k+1} - v_2 \hat{d}_{2,x}^{k+1} \\
&\quad - \lambda_2 \hat{d}_2^{k+1} + \lambda_1 \Psi \text{ over } \Omega_2, \quad t \in [T_0, T_f], \\
\hat{d}_2^{k+1}(L_1, t) &= ||e_2^k(L_1, t)||_\infty, \quad t \in [[T_0, T_f], \\
\hat{d}_2^{k+1}(L, t) &= 0, \quad t \in [T_0, T_f], \\
\hat{d}_2^{k+1}(x, t_0) &= \mathcal{B}(x), \quad x \in \Omega_2,
\end{aligned} \qquad (5.84)$$

where

$$\mathcal{B}(x) = ||e^k(L_1, t)||_\infty e^{\alpha_2(x-L_1)} \frac{\sinh(\beta_2(x-L))}{\sinh(\beta_2(L_1-L))}$$

$$+ \frac{\lambda_1}{\lambda_2} \Psi \frac{\sinh(\beta_2(L-x))}{\sinh(\beta_2(L_1-L))} \left[e^{\alpha_2(x-L_1)} - e^{\alpha_2(x-L)} e^{\beta_2(L-L_1)} \right] \qquad (5.85)$$

$$- \frac{\lambda_1}{\lambda_2} \Psi \left[1 - e^{\alpha_2(x-L)} e^{\beta_2(L-x)} \right].$$

Then the solution to (5.82) and (5.84) is the steady state solution given by

$$\hat{e}_2^{k+1}(x) = ||d_2^k(L_2, t)||_\infty e^{\alpha_2(x-L_2)} \frac{\sinh(\beta_2 x)}{\sinh(\beta_2 L)}$$

$$+ \frac{\lambda_1}{\lambda_2} \Psi \left[e^{\alpha_2 x} \frac{\sinh(\beta_2(x-L_2))}{\sinh(\beta_2 L_2)} - e^{\alpha_2(x-L_2)} \frac{\sinh \beta_2 x}{\sinh \beta_2 L_2} + 1 \right],$$

and

$$\hat{d}_2^{k+1}(x) = ||e^k(L_1, t)||_\infty e^{\alpha_2(x-L_1)} \frac{\sinh(\beta_2(x-L))}{\sinh(\beta_2(L_1-L))}$$

$$+ \frac{\lambda_1}{\lambda_2} \Psi \frac{\sinh(\beta_2(L-x))}{\sinh(\beta_2(L_1-L))} \left[e^{\alpha_2(x-L_1)} - e^{\alpha_2(x-L)} e^{\beta_2(L-L_1)} \right]$$

$$- \frac{\lambda_1}{\lambda_2} \Psi \left[1 - e^{\alpha_2(x-L)} e^{\beta_2(L-x)} \right],$$

respectively.

By defining the function $E(x, t) = \hat{e}^{k+1} - e^{k+1}$, as in the proof of Theorem 5.2, and by the maximum principle theorem we conclude that

$$|e_2^{k+1}| \leq \hat{e}_2^{k+1}$$

for all (x, t), and similarly,

$$|d_2^{k+1}| \leq \hat{d}_2^{k+1}.$$

Then

$$|e_2^{k+1}(x,t)| \leq ||d_2^k(L_2,t)||_\infty e^{\alpha_2(x-L_2)} \frac{\sinh(\beta_2 x)}{\sinh(\beta_2 L)}$$

$$+ \tfrac{\lambda_1}{\lambda_2} \Psi \left[e^{\alpha_2 x} \frac{\sinh(\beta_2(x-L_2))}{\sinh(\beta_2 L_2)} - e^{\alpha_2(x-L_2)} \frac{\sinh \beta_2 x}{\sinh \beta_2 L_2} + 1 \right], \qquad (5.86)$$

and

$$|d_2^{k+1}(x,t)| \leq ||e^k(L_1,t)||_\infty e^{\alpha_2(x-L_1)} \frac{\sinh(\beta_2(x-L))}{\sinh(\beta_2(L_1-L))}$$

$$+ \tfrac{\lambda_1}{\lambda_2} \Psi \frac{\sinh(\beta_2(L-x))}{\sinh(\beta_2(L_1-L))} \left[e^{\alpha_2(x-L_1)} - e^{\alpha_2(x-L)} e^{\beta_2(L-L_1)} \right] \qquad (5.87)$$

$$- \tfrac{\lambda_1}{\lambda_2} \Psi \left[1 - e^{\alpha_2(x-L)} e^{\beta_2(L-x)} \right].$$

By evaluating (5.87) for $d_2^k(x,t)$ at $x = L_2$, substituting the results in (5.86) and afterwards evaluating the resulting relation at $x = L_1$, we observe that (5.78) holds in general.

Similarly for (5.79) which will follow from the evaluation of $e_2^{k+1}(x,t)$ at $x = L_1$, substituting in (5.87) and followed by evaluating the resulting relation at $x = L_2$. □

For the decoupled case of the convection-diffusion-reaction-equation, the convergence of the overlapping Schwarz waveform-relaxation method given by Theorem 5.3 depends on the factor $\gamma_i = \frac{\sinh(\beta_i L_1)}{\sinh(\beta_i L_2)} \frac{\sinh \beta_i(L_2-L)}{\sinh \beta_i(L_1-L)}$ which is concluded earlier for single scalar convection-diffusion-reaction by Theorem 5.2.

For the coupled system we illustrated Theorem 5.4 and assume that the error depends on two main factors: the convergence parameter γ_i and the coupling parameter λ_1 defining the coupled system (5.70), (5.71). It is obvious that for the coupling parameter $\lambda_1 = 0$, we retain the decoupled system and a faster convergence rate is achieved if we have a small ratio $\frac{\lambda_1}{\lambda_2}$.

Chapter 6

Numerical Experiments

6.1 Introduction

In the numerical experiments we discuss test examples and real-life applications with respect to our proposed decomposition methods. We present complex models and their tendency to arrive at inefficient and inexact solutions, and we are forced to search for more detailed and exact solutions for the same problems using simpler equations.

In the future, we will propose more solutions that are adequate in their results for multi-physics applications rather than analytical exact solutions, which involve complex equations. Our solutions offer the possibility of mathematically exact proofs of existence and uniqueness, whereas the older computations of greater complexity lack these convergence and existence results.

We will defer to mathematical correctness if there is a chance to fulfill this in the simpler equations, but we will also describe very complex models and show solvability without proofs of existence and uniqueness. To find a balance between simple provable equations and complex calculable equations, we present splitting methods for decoupling complex equations into provable equations.

Complex models will be described with more or less understanding of the complexity of the particular systems. Therefore, systematic schemes are used to decouple problems in simpler understandable models and, for the next step, to couple in more complex models. In this way, understanding the part-systems is possible, and the complex model is made at least partially understandable. In this chapter, we introduce various models in solid and fluid mechanics with their physical background, and we discuss mathematical proofs for solutions to physical problems in various situations.

6.2 Benchmark Problems for the Time Decomposition Methods for Ordinary Differential and Parabolic Equations

For the qualitative characterization of the time decomposition methods, we introduce in the following the benchmark problems.

We have chosen model problems, in which the exact solutions are known, so that we can define the exact values of the errors.

In our examples, we first consider a simple scalar equation as ordinary differential equation (ODE), and after that, we consider systems of ODEs and parabolic equations. We present the flexibility and improvement of the iterative operator-splitting method. In the scheme of the various operator-splitting methods, we also use the analytical method of such reduced ODEs and parabolic equations. We can verify the number of iteration steps with respect to the order of the approximation of the functions.

6.2.1 First Test Example: Scalar Equation

In the following, we introduce the application of the splitting methods on a simple scalar equation, which can be decoupled and solved as exact as the nonsplitted version.

Because of the iterative method, we have to investigate the computational time for this smoothing method. This we have to take into account for decomposing a simple standard example.

We consider the following Cauchy problem for the scalar equation:

$$\frac{du(t)}{dt} = (-\lambda_1 - \lambda_2)u(t), \ t \in [0, T], \tag{6.1}$$

$$u(0) = u_0, \tag{6.2}$$

which has the exact solution

$$u(t) = \exp(-(\lambda_1 + \lambda_2))t)u_0. \tag{6.3}$$

For the problem (6.1) we split the right-hand side into the sum of two scalar operators $A + B$, where $Au = -\lambda_1 u$ and $Bu = -\lambda_2 u$. According to the iterative splitting method (see Chapter 3), we apply the following algorithm:

$$\frac{du_i(t)}{dt} = -\lambda_1 u_i(t) - \lambda_2 u_{i-1}(t), \tag{6.4}$$

$$\frac{du_{i+1}(t)}{dt} = -\lambda_1 u_i(t) - \lambda_2 u_{i+1}(t), \tag{6.5}$$

on the interval $t \in [0, T]$, where $u_i(0) = u_{i+1}(0) = u_0$, $u_0(t) = 0$, $\forall t \in [0, T]$, and $i = 1, 3, 5, \ldots, 2m + 1$ is the number of iterations, with m as a positive integer.

For the two equations (6.4) and (6.5), we can derive the analytical solutions as

$$u_i(t) = \exp(-\lambda_1 t)\, u_i(0) + \frac{\lambda_2}{\lambda_1} u_{\text{approx},i-1}(t)\, (\exp(-\lambda_1 t) - 1), \quad (6.6)$$

$$u_{i+1}(t) = \frac{\lambda_1}{\lambda_2} u_{\text{approx},i}(t)\, (\exp(-\lambda_2 t) - 1) + \exp(-\lambda_2 t)\, u_{i+1}(0), \quad (6.7)$$

where the initial conditions are $u_{i+1}(0) = u_0$ and $u_i(0) = u_0$ with the index $i = 1, 3, 5, \ldots, 2m + 1$. The time interval is $t \in [0, T]$. The starting solutions are fixed as $u_0(t) = 0$. Further, $u_{\text{approx},i-1}$ are the approximated solutions for the last iterative solution u_{i-1}, which has at least an accuracy of $\mathcal{O}(\tau^{2m+1})$ (with τ is the time-step).

Based on these solutions, we compare the results of the iterative splitting method with the analytical solution of the complex equation, see also [86].

We perform the time discretization with the trapezoidal rule, which is a second-order method.

The combination by handling both the iteration steps and the time partitions is therefore important. We assume a time interval $[0, T]$ and divide it in n intervals with the length $\tau_n = \frac{T}{n}$. We could improve the results by using smaller time-steps and more iteration steps. The optimal relation is an adequately large time-step with fewer iteration steps. Because of the approximation of our initial function, we can conclude that two to four iteration steps are sufficient, cf. Theorem 3.1.

For our example we chose $\lambda_1 = 0.25$, $\lambda_2 = 0.5$, and $T = 1.0$, such that we obtain our exact solution with $u_{\text{exact}} = \exp(-0.75) \approx 0.4723665$.

For the simulation we use MATLAB 7.0 on a single CPU (2.1 GHZ).

In Table 6.1 we have the errors at time $T = 1.0$ between the analytical and numerical results, and the computational time in $[sec]$, for the nonsplitting method. In the Table 6.2 we have the errors at time $T = 1.0$ and the computational time in $[sec]$, for the splitting method.

TABLE 6.1: Numerical results for the first example with nonsplitting method and second-order trapezoidal rule.

Number of time partitions n	err $= \|u_{\text{exact}} - u_{\text{num}}\|$	Comput. time [sec]
100	1.5568e-12	1.82e-02
1000	1.0547e-15	1.41e-01

For the splitting method we obtain for few time partitions and much iteration steps the best results, see $n = 1$ and $i = 100$, but we have to deal with

TABLE 6.2: Numerical results for the first example with splitting method and second-order trapezoidal rule.

Number of time partitions n	Number of iterations i	err = $|u_{\text{exact}} - u_{\text{num}}|$	Comput. time [sec]
1	2	3.8106e-02	2.83e-01
1	4	4.1633e-04	5.41e-01
1	10	5.5929e-12	1.24e+00
1	100	9.4369e-16	1.22e+01
5	2	6.1761e-03	2.80e-01
5	4	2.6127e-06	5.48e-01
5	10	1.0547e-15	1.32e+00
10	2	3.0185e-03	3.01e-01
10	4	3.1691e-07	5.85e-01
10	10	1.0547e-15	1.42e+00
100	2	2.9588e-04	6.77e-01
100	4	3.0845e-10	1.31e+00
100	10	6.6613e-16	3.21e+00

a large amount of computational time. We suggest to optimize with respect to the computational time, and in comparison to the nonsplitting method, to propose more time partitions and fewer iterations (e.g., $i = 2$ and $n = 10$). Comparing our theoretical convergence results, we obtain a convergence order of 1 more for each iteration step, so at least more iteration steps result in more accurate solutions.

REMARK 6.1 In this example, we show the additional amount of work for the iterative method. It presents that the splitting method is as accurate as the nonsplitting method. With respect to optimizing the number of time partitions and iterations, we can perform the splitting method. The acceleration can also be done with parallel computing, with respect to more iterations; therefore, the splitting method is as attractive as the nonsplitting methods for such examples.   ⬚

6.2.2 Second Test Example of a System of ODEs: Stiff and Nonstiff Case

Let us consider a more complicated example of these computations, where the motivation is a reversible chemical reaction process between two species. We could apply the example to chemical reaction models and the bio remediation of complex processes, cf. [80].

6.2.2.1 Linear ODE with Nonstiff Parameters

In the first example, we deal with the following linear ordinary differential equation:

$$\frac{\partial u(t)}{\partial t} = \begin{pmatrix} -\lambda_1 & \lambda_2 \\ \lambda_1 & -\lambda_2 \end{pmatrix} u(t), \ t \in [0, T], \tag{6.8}$$

$$u(0) = u_0, \tag{6.9}$$

where the initial condition $u_0 = (1, 1)$ is given on the interval $[0, T]$.

The analytical solution is given by

$$u(t) = \begin{pmatrix} c_1 - c_2 \exp\left(-(\lambda_1 + \lambda_2)t\right) \\ \frac{\lambda_1}{\lambda_2} c_1 + c_2 \exp\left(-(\lambda_1 + \lambda_2)t\right) \end{pmatrix}, \tag{6.10}$$

where

$$c_1 = \frac{2}{1 + \frac{\lambda_1}{\lambda_2}}, \quad c_2 = \frac{1 - \frac{\lambda_1}{\lambda_2}}{1 + \frac{\lambda_1}{\lambda_2}}. \tag{6.11}$$

We split our linear operator into two operators by setting

$$\frac{\partial u(t)}{\partial t} = \begin{pmatrix} -\lambda_1 & 0 \\ \lambda_1 & 0 \end{pmatrix} u(t) + \begin{pmatrix} 0 & \lambda_2 \\ 0 & -\lambda_2 \end{pmatrix} u(t). \tag{6.12}$$

We choose $\lambda_1 = 0.25$ and $\lambda_2 = 0.5$ on the interval $[0, 1]$ (i.e., with $T = 1$).

We therefore have the following operators:

$$A = \begin{pmatrix} -0.25 & 0 \\ 0.25 & 0 \end{pmatrix}, \quad B = \begin{pmatrix} 0 & 0.5 \\ 0 & -0.5 \end{pmatrix}. \tag{6.13}$$

For our time integration method, we assume a time interval $[0, T]$ and divide it in n intervals with the length $\tau_n = \frac{T}{n}$. We can improve our results by using smaller time-steps and more iteration steps.

For the initialization of our iterative method, for $i = 1$, we use $u_0(0) = (0, 0)^t$. From the examples, one can see that the order increases by each iteration step.

In the following, we compare the results of different discretization methods for a linear ordinary differential equation. For the time discretization we apply the second-order trapezoidal rule, the third-order backward differentiation formula (BDF3), see [116], and at least a fourth-order Gauss-Runge-Kutta method (Gauss-RK), see [33], [34], and [116]. Thus, we can obtain a maximal fourth-order method with our iterative operator-splitting method.

For the simulation we use MATLAB 7.0 on a single Linux-PC with $2.1GHz$.

In Table 6.3 we have the errors between the analytical and numerical results, and the computational time in $[sec]$, for the nonsplitting method.

Our numerical results for the splitting method are presented in Tables 6.4, 6.5, and 6.6. Additionally we present in Table 6.4 the numerical results and computational time in $[sec]$ for the nonsplitting method.

To compare the results we chose the same number of iteration steps and time partitions for the splitting methods. The error between the analytical and numerical solution is given in the supremum norm that is, $\text{err}_k = |u_{k,\text{exact}} - u_{k,\text{num}}|$, with $k = 1, 2$.

TABLE 6.3:   Numerical results for the second example with nonsplitting method and second-order trapezoidal rule.

Number of splitting partitions n	err_1	err_2	Comput. time [sec]
100	5.1847e-13	5.1903e-13	2.27e-02
1000	2.2204e-15	3.3307e-16	1.92e-01

TABLE 6.4:   Numerical results for the second example with iterative splitting method and second-order trapezoidal rule.

Iteration steps i	Number of splitting partitions n	err_1	err_2	Comput. time [sec]
2	1	4.5321e-002	4.5321e-002	3.51e-01
2	10	3.9664e-003	3.9664e-003	3.93e-01
2	100	3.9204e-004	3.9204e-004	7.83e-01
3	1	7.6766e-003	7.6766e-003	5.15e-01
3	10	6.6383e-005	6.6383e-005	5.76e-01
3	100	6.5139e-007	6.5139e-007	1.15e+00
4	1	4.6126e-004	4.6126e-004	6.85e-01
4	10	4.1321e-07	4.1321e-07	7.63e-01
4	100	4.0839e-10	4.0839e-10	1.52e+00
5	1	4.6828e-005	4.6828e-005	8.51e-01
5	10	4.1382e-09	4.1382e-09	9.47e-01
5	100	4.0878e-13	4.0856e-13	1.89e+00
6	1	1.9096e-006	1.9096e-006	1.02e+00
6	10	1.7200e-11	1.7200e-11	1.13e+00
6	100	2.4425e-15	1.1102e-16	2.25e+00

TABLE 6.5: Numerical results for the second example with iterative splitting method and BDF3 method.

Iteration steps i	Number of splitting partitions n	err_1	err_2
2	1	4.5321e-002	4.5321e-002
2	10	3.9664e-003	3.9664e-003
2	100	3.9204e-004	3.9204e-004
3	1	7.6766e-003	7.6766e-003
3	10	6.6385e-005	6.6385e-005
3	100	6.5312e-007	6.5312e-007
4	1	4.6126e-004	4.6126e-004
4	10	4.1334e-007	4.1334e-007
4	100	1.7864e-009	1.7863e-009
5	1	4.6833e-005	4.6833e-005
5	10	4.0122e-009	4.0122e-009
5	100	1.3737e-009	1.3737e-009
6	1	1.9040e-006	1.9040e-006
6	10	1.4350e-010	1.4336e-010
6	100	1.3742e-009	1.3741e-014

A higher order in the time discretization allows improved results with more iteration steps. Based on the theoretical results, we can improve the order of the results with each iteration step. Thus, for the fourth-order time discretization, we can show the highest order in our iterative method. In comparison to the nonsplitting method, with only applying the time discretization methods, we can reach the same accuracy, but we have to take into account the increased amount of computational time. An optimal choice of fewer iterations and sufficient time partitions can save the resources, and we can propose it as a competitive method.

The convergence results of the three methods are given in Figure 6.1.

REMARK 6.2 For the nonstiff case, we obtain improved results for the iterative splitting method by increasing the number of iteration steps. Due to improved time discretization methods, the splitting error can be reduced with higher-order Runge-Kutta methods. Further, the optimal choice of iterations and sufficient time partitions can make the splitting method more attractive to such standard problems. By the way, the application of such iterative splitting methods becomes more attractive for complicated equations, while saving memory and computational time for the appropriate decomposition. □

TABLE 6.6: Numerical results for the second example with iterative splitting method and fourth-order Gauß-RK method.

Iteration steps i	Number of splitting partitions n	err_1	err_2
2	1	4.5321e-002	4.5321e-002
2	10	3.9664e-003	3.9664e-003
2	100	3.9204e-004	3.9204e-004
3	1	7.6766e-003	7.6766e-003
3	10	6.6385e-005	6.6385e-005
3	100	6.5369e-007	6.5369e-007
4	1	4.6126e-004	4.6126e-004
4	10	4.1321e-007	4.1321e-007
4	100	4.0839e-010	4.0839e-010
5	1	4.6833e-005	4.6833e-005
5	10	4.1382e-009	4.1382e-009
5	100	4.0878e-013	4.0856e-013
6	1	1.9040e-006	1.9040e-006
6	10	1.7200e-011	1.7200e-011
6	100	2.4425e-015	1.1102e-016

6.2.2.2 Linear ODE with Stiff Parameters

We deal with the same equation as in the first example, now choosing $\lambda_1 = 1$ and $\lambda_2 = 10^4$ on the interval $[0,1]$.

We therefore have the following operators:

$$A = \begin{pmatrix} -1 & 0 \\ 1 & 0 \end{pmatrix} \quad , \quad B = \begin{pmatrix} 0 & 10^4 \\ 0 & -10^4 \end{pmatrix}. \qquad (6.14)$$

The discretization of the linear ordinary differential equation is done with the BDF3 method. Our numerical results are presented in Table 6.7. For the stiff problem, we choose more iteration steps and time partitions and show the error between the analytical and numerical solution in the supremum norm that is, $err_k = |u_{k,\text{exact}} - u_{k,\text{num}}|$ with $k = 1, 2$.

In Table 6.7, we see that we must double the number of iteration steps to obtain the same results for the nonstiff case.

REMARK 6.3 For the stiff case, we obtain improved results with more than five iteration steps. Because of the inexact starting function, the accuracy must be improved by more iteration steps. Higher-order time discretization methods, such as BDF3 method and iterative operator-splitting method, accelerate the solving process. ⬚

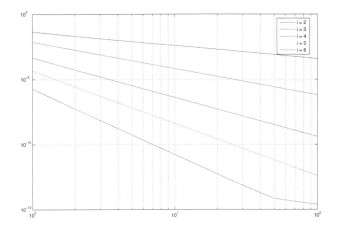

FIGURE 6.1: Convergence rates from two up to six iterations.

6.2.3 Third Example: Linear Partial Differential Equation

We consider the one-dimensional convection-diffusion-reaction equation given by

$$R\partial_t u + v\partial_x u - D\partial_{xx} u = -\lambda u \ , \ \text{in } \Omega \times [t_0, t_{\text{end}}), \tag{6.15}$$

$$u(x, t_0) = u_{\text{exact}}(x, t_0) \ , \ \text{in } \Omega, \tag{6.16}$$

$$u(0, t) = u_{\text{exact}}(0, t) \ , \ u(L, t) = u_{\text{exact}}(L, t). \tag{6.17}$$

We choose $x \in [0, L] = \Omega$ with $L = 30$ and $t \in [t_0, t_{\text{end}}] = [10^4, 2 \cdot 10^4]$. Furthermore, we have $\lambda = 10^{-5}$, $v = 0.001$, $D = 0.0001$, and $R = 1.0$. The analytical solution is given by

$$u_{\text{exact}}(x, t) = \frac{1}{2\sqrt{D\pi t}} \exp(-\frac{(x - vt)^2}{4Dt}) \exp(-\lambda t). \tag{6.18}$$

To begin away from the singular point of the exact solution, we start from the time point $t_0 = 10^4$.

Our split operators are

$$A = \frac{D}{R}\partial_{xx}, \ B = -\frac{1}{R}(\lambda + v\partial_x). \tag{6.19}$$

For the spatial discretization we use the finite differences with a spatial grid width of $\Delta x = \frac{1}{10}$ and $\Delta x = \frac{1}{100}$. The discretization of the linear ordinary differential equation is accomplished with the BDF3 method, so we are dealing with a third-order method. Our numerical results are presented in Table 6.8.

TABLE 6.7: Numerical results for the stiff example with iterative operator-splitting method and BDF3 method.

Iteration steps i	Number of splitting partitions n	err_1	err_2
5	1	3.4434e-001	3.4434e-001
5	10	3.0907e-004	3.0907e-004
10	1	2.2600e-006	2.2600e-006
10	10	1.5397e-011	1.5397e-011
15	1	9.3025e-005	9.3025e-005
15	10	5.3002e-013	5.4205e-013
20	1	1.2262e-010	1.2260e-010
20	10	2.2204e-014	2.2768e-018

TABLE 6.8: Numerical results for the third example with iterative operator-splitting method and BDF3 method with $\Delta x = 10^{-2}$.

Iteration steps i	Number of splitting partitions n	err at $x = 18$	err at $x = 20$	err at $x = 22$
1	10	9.8993e-002	1.6331e-001	9.9054e-002
2	10	9.5011e-003	1.6800e-002	8.0857e-003
3	10	9.6209e-004	1.9782e-002	2.2922e-004
4	10	8.7208e-004	1.7100e-002	1.5168e-005

We choose different iteration steps and time partitions, and show the error between the analytical and numerical solution in the supremum norm.

Figure 6.2 shows the initial solution at $t_0 = 10^4$, and the analytical as well as the numerical solutions at $t_{\text{end}} = 2 \cdot 10^4$ of the convection-diffusion-reaction equation.

One result we can see is that we can reduce the error between the analytical and the numerical solution by using more iteration steps. If we restrict ourselves to an error of 10^{-4}, we obtain an effective computation with three iteration steps and ten time partitions.

REMARK 6.4 For the partial differential equations, we also need to take into account the spatial discretization. We applied a fine grid step on the spatial discretization, such that the error of the time discretization method is dominant. We obtain an optimal efficiency of the iteration steps and the time partitions, if we use ten iteration steps and two time partitions. □

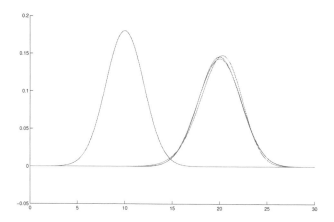

FIGURE 6.2: Initial and computed results for the third example with iterative splitting method and BDF3 method; the initial solution is plotted as blue line, the iterative solution is plotted as green line, and the exact solution is plotted as red line.

6.2.4 Fourth Example: Nonlinear Ordinary Differential Equation

As a nonlinear differential example, we choose the Bernoulli equation:

$$\frac{\partial u(t)}{\partial t} = \lambda_1 u(t) + \lambda_2 (u(t))^p, \ t \in [0, T], \tag{6.20}$$

$$u(0) = 1, \tag{6.21}$$

with the solution

$$u(t) = \left[\left(1 + \frac{\lambda_2}{\lambda_1} \right) \exp(-\lambda_1 t (1 - p)) - \frac{\lambda_2}{\lambda_1} \right) \right]^{\frac{1}{1-p}}, \text{ with } p \neq 0, 1. \tag{6.22}$$

We choose $p = 2$, $\lambda_1 = -1$, $\lambda_2 = -100$, and $\Delta x = 10^{-2}$. For the time interval $[0, T]$, with $T = 1$, we divide in n intervals with the length $\tau_n = \frac{T}{n}$.

We apply the iterative operator-splitting method with the nonlinear operators

$$A(u) = \lambda_1 u(t), \ B(u) = \lambda_2 (u(t))^p. \tag{6.23}$$

The discretization of the nonlinear ordinary differential equation is performed with higher-order Runge-Kutta methods, precisely at least third-order methods, see also [87]. Our numerical results are presented in Table 6.9. We choose different iteration steps and time partitions. The error between the analytical and numerical solution is shown with the supremum norm at time

TABLE 6.9: Numerical results for the Bernoulli equation with iterative operator-splitting method and BDF3 method.

Iteration steps i	Number of splitting partitions n	err
2	1	7.3724e-001
2	2	2.7910e-002
2	5	2.1306e-003
10	1	1.0578e-001
10	2	3.9777e-004
20	1	1.2081e-004
20	2	3.9782e-005

$T = 1.0$. The experiments result in showing the reduced errors for more iteration steps and more time partitions. Because of the time discretization method for ODEs, we restrict the number of iteration steps to a maximum of five. If we restrict the error bound to 10^{-3}, two iteration steps and five time partitions give the most effective combination.

REMARK 6.5 For nonlinear ordinary differential equations, the exact starting function is a problem. Therefore, the initialization process is delicate and we can decrease the splitting error by more iteration steps. Due to the linearization, we gain almost linear convergence rates. This can be improved by a higher-order linearization, see [16], [155]. ▯

6.2.5 Fifth Example: Partial Differential Equation

In this example, we simulate a general single-species reactive transport equation, which can be written as

$$u_t + \nabla \cdot (v\,u - D\,\nabla u) = R(u) + f(x,t), \text{ in } \Omega \times (0,T), \qquad (6.24)$$
$$u(x,t) = 0, \text{ on } \partial\Omega \times (0,T), \qquad (6.25)$$
$$u(x,0) = u_0(x), \text{ on } \Omega, \qquad (6.26)$$

where $[0,T]$ is a time interval, $x = (x_1, \ldots, x_d)^T \in \Omega$ is the space variable, and Ω is a domain in $\mathbb{R}^d$ $(d = 1, 2, \text{ or}, 3)$. $u(x,t)$ is the unknown population density or concentration of the species, $v \in \mathbb{R}^{d,+}$ is a divergence-free velocity field, $D \in \mathbb{R}^{d,+} \times \mathbb{R}^{d,+}$ is a diffusion tensor (it is assumed that elementwise $|D_{i,j}| << |v_i| \,\forall i, j \in \{1, \ldots, d\}$), $R(u)$ is a nonlinear reaction term, and $u_0(x)$ is an initial condition.

For the reaction term, we are concerned with the

- Radioactive decay with $R(u) = -au$, as a linear reaction term

- Bio degradation model with $R(u) = au/(u+b)$, as a nonlinear reaction term

The models and detailed discussions on bio degradation can be found in [40].

We propose, based on the discretization methods and the underlying timescales, that the convection-reaction part of the equation serves as the basis of one operator and that the diffusion part and the right-hand side serve as the basis of a second operator. Because of the splitting, the two equations can be dealt with separately with most effective discretization methods in each case. We propose for the convection-reaction part characteristic methods and a finite element method for the diffusion part.

The transport equation (6.24) can be rewritten as an abstract operator equation:

$$u_t = A(u) + B(u), \tag{6.27}$$

with

$$A(u) = -v \cdot \nabla u + R(u), \tag{6.28}$$
$$B(u) = \nabla \cdot (D\nabla u) + f(x,t). \tag{6.29}$$

Let N be a positive integer, $\tau_n := T/N$ as time-step, and let $t_n = n\tau_n$ ($n = 0, 1, \cdots, N$) be a uniform partition of the time period $[0, T]$. We split the equation (6.24) into two equations in each small time period $[t_n, t_{n+1}]$:

$$u_t + v \cdot \nabla u = R(u), \tag{6.30}$$

$$u_t = \nabla \cdot (D\nabla u) + f(x,t). \tag{6.31}$$

For any $x \in \Omega$, the characteristic $y(s; x, t_{n+1})$ passing through (x, t_{n+1}) is determined by

$$\begin{cases} \dfrac{dy}{ds} = v(y, s), & s \in (t_n, t_{n+1}), \\ y(t_{n+1}; x, t_{n+1}) = x. \end{cases} \tag{6.32}$$

Let (x^*, t_n) be the image of exact backtracking of (x, t_{n+1}). If the characteristic reaches the boundary, then we use (x^*, t_n^*), where $x^* \in \partial\Omega$, $t_n \le t_n^* \le t_{n+1}$. Notice that exact tracking of characteristics is usually unavailable in practice and we have to resort to numerical means. All commonly used numerical methods for solving ODEs (e.g., Euler and Runge-Kutta methods), can be applied to problem (6.32).

Along characteristics, the convection-reaction equation becomes a nonlinear ODE:

$$\begin{cases} \dfrac{du^{(1)}}{dt} = R(u^{(1)}), & t \in (t_n^*, t_{n+1}), \\ u^{(1)}(x^*, t_n^*) = u^{(2)}(x^*, t_n^*), \end{cases} \tag{6.33}$$

where $u^{(2)}$ is the solution for the parabolic problem and will be explained later. When $n = 0$ or $x^* \in \partial\Omega$, we should replace $u^{(2)}(x^*, t_n^*)$ by $u_0(x^*)$ or the boundary condition.

Problem (6.33) can be solved numerically by for example, Euler and Runge-Kutta methods. With a numerical solution, the nonlinearity in the reaction term can be well resolved if $R(u)$ is Lipschitz continuous with respect to u, which is true for the bio degradation models and also logistic models, see [94].

The other part of the equation to solve is an initial boundary value problem for a typical parabolic equation

$$
\begin{cases}
u_t^{(2)} = \nabla \cdot (D\nabla u^{(2)}) + f(x,t), & x \in \Omega, \ t \in (t_n, t_{n+1}), \\
u^{(2)}|_{\partial\Omega} = 0, \\
u^{(2)}(x, t_n) = u^{(1)}(x, t_{n+1}).
\end{cases}
\tag{6.34}
$$

This problem is conventional and can be solved by finite difference, element, or volume methods.

Let Δx be the spatial mesh size in the numerical scheme (finite difference, element, or volume) for solving the parabolic part (6.34), τ is the temporal step size. Because $|D| << 1$, the stability condition $|D|\tau/\Delta x^2 \leq 1/2$ is readily satisfied.

6.2.5.1 Linear Reaction

To examine our method, we first consider a two-dimensional problem with a linear reaction, to which we can find the exact solution. This enables us to compare the numerical and exact solutions. In particular, we have a rotating velocity $v = (-4y, 4x)$, a constant scalar diffusion $D > 0$, a linear reaction $R(u) = Ku$ with K being a constant, and a null source/sink (i.e., $f \equiv 0$). Assume the substance is initially normally distributed that is, the initial condition is specified as a Gaussian hill:

$$
u_0(x, y) = \exp\left(-\frac{(x - x_c)^2 + (y - y_c)^2}{2\sigma^2}\right),
\tag{6.35}
$$

where (x_c, y_c) is the center and $\sigma > 0$ is the standard deviation. Then the exact solution is given by

$$
u(x, y, t) = \frac{2\sigma^2}{2\sigma^2 + 4Dt} \exp\left(Kt - \frac{(x^* - x_c)^2 + (y^* - y_c)^2)}{2\sigma^2 + 4Dt}\right),
\tag{6.36}
$$

where $(x^*, y^*, 0)$ is the backtracking foot of the characteristic from (x, y, t) that is,

$$
\begin{cases}
x^* = (\cos 4t)x + (\sin 4t)y, \\
y^* = -(\sin 4t)x + (\cos 4t)y.
\end{cases}
\tag{6.37}
$$

TABLE 6.10: Numerical results of example with linear reaction and $\tau = \pi/8$.

Mesh size Δx	err_{u,L_∞}	err_{u,L_1}	err_{u,L_2}
1/20	1.266×10^{-2}	1.247×10^{-4}	3.138×10^{-4}
1/40	1.031×10^{-2}	5.061×10^{-5}	2.085×10^{-4}
1/50	9.984×10^{-3}	4.153×10^{-5}	1.923×10^{-4}
1/60	9.796×10^{-3}	3.613×10^{-5}	1.825×10^{-4}

For simplicity, we use a uniform triangular mesh. The second-order Runge-Kutta (or Heun) method is used for characteristic tracking even though exact tracking is available. The finite element solver for the parabolic part can be implemented as a dynamic link library (DLL) which is derived from the source code in an object finite element library (OFELI), see [162]. Therefore, the implementation is at least only a recoding and application of established FEM libraries and can be programmed very quickly.

For numerical runs, we choose $T = \pi/2$, $\Omega = [-1,1] \times [-1,1]$, $D = 10^{-4}$, $K = 0.1$, $(x_c, y_c) = (-0.5, -0.5)$, and $\sigma^2 = 0.01$. For the parabolic solver, we use 20 micro steps within each global time-step $[t_n, t_{n+1}]$. Accordingly, we set the maximum number of time-steps in the characteristic tracking to 20. Table 6.10 lists some results for the numerical solution at the final time. We still obtain very good numerical solutions even with large global time-steps.

6.2.5.2 Nonlinear Reaction

The second example is a simplified model for single-species bio degradation: $R(u) = au/(u + b)$. We consider a two-dimensional problem with a constant velocity field $(v_1, v_2)^t$, a scalar diffusion $D > 0$, and no source/sink. The initial condition is a normal distribution (Gaussian hill). Again we use the parabolic solver (DLL) compiled from OFELI.

TABLE 6.11: Numerical results of example with nonlinear reaction.

time-step τ	Mesh size Δx	u_{min}	u_{max}
0.25	1/20	0.0	1.5159
0.25	1/40	0.0	1.5176
0.25	1/60	0.0	1.5179
0.125	1/40	0.0	1.5251
0.10	1/20	0.0	1.5248
0.10	1/50	0.0	1.5268

For numerical runs, we choose $T = 1$, $\Omega = [-1,1] \times [-1,1]$, $(v_1, v_2)^t = (1,1)$,

$D = 10^{-4}$, $a = b = 1$, $(x_c, y_c) = (-0.5, -0.5)$, and $\sigma^2 = 0.01$. Table 6.11 lists some results of the numerical solution at the final time.

No exact solution is known for this problem. However, from Table 6.11, we observe that the operator-splitting method is stable and keeps positivity of the solution.

REMARK 6.6 We apply an operator-splitting method for transport equations with nonlinear reactions. In the implementation of the proposed method, we incorporate some existing commercial and free software components. By integrating functionalities of existing software components, application developers do not have to create everything from scratch.

In the numerical results, we see that the method works very well and that we have efficient computations. A simple extension of the proposed operator-splitting method can also be made to coupled systems, see [105].

□

6.3 Benchmark Problems for the Spatial Decomposition Methods: Schwarz Waveform-Relaxation Method

In this section, we will present the benchmark problems for the spatial decomposition method. They incorporate operator-splitting methods as well as Schwarz waveform-relaxation methods.

6.3.1 First Example: Convection-Diffusion-Reaction Equation

The problems are discretized using second-order approximations with respect to the spatial variable using a regular spatial mesh with $\Delta x = L/N$, and backward approximations with respect to the time using τ as time-step. The first-order operator-splitting method (FOP) is considered to be the basic solution algorithm for the overlapping Schwarz waveform-relaxation method (FOPSWR). We also consider the second-order operator-splitting method (SOP), performed as the Strang splitting method, see [185], and considered to be the basic solution algorithm for the overlapping Schwarz waveform-relaxation method (SOPSWR).

We consider the one-dimensional convection-diffusion-reaction equation given as

$$\partial_t u + v \partial_x u - \partial_x D \partial_x u = -\lambda u, \tag{6.38}$$

$$u(x, t_0) = u_{exact}(x, t_0), \quad u(\pm\infty, t) = 0, \tag{6.39}$$

defined over $\Omega \times [t_0, T_f] = \Omega \times [100, 10^5]$, with an exact solution given as

$$u_{\text{exact}}(x, t) = \frac{u_0}{2\sqrt{D\pi t}} \exp(-\frac{(x - vt)^2}{4Dt}) \exp(-\lambda t). \qquad (6.40)$$

The initial condition and the Dirichlet boundary condition are defined using the exact solution (6.40) at the start time $t_0 = 100$ and with $u_0 = 1.0$. The reaction parameter is given as $\lambda = 10^{-5}$, where the convection parameter is $v = 0.001$ and the diffusion parameter is $D = 0.0001$. The time interval is given as $[100, 10^5]$. The boundary condition is given by $u(\pm\infty, t) = 0$.

We considered the backward Euler discretization for both splitted operators (i.e., the convection and the diffusion reaction operator), to simulate the solution over the time interval $[100, 10^5]$.

The model problem (6.38) is solved with both first-order operator-splitting method (FOP) and first-order operator-splitting method with overlapping Schwarz waveform-relaxation method (FOPSWR).

We compare the accuracy of the solution over the entire spatial domain with different values for the spatial step Δx, and equidistant time-steps τ using both the FOP method and the FOPSWR method over two subdomains with different sizes of overlapping. The errors of the solution are given in Table 6.12 and Table 6.13, respectively. The FOPSWR method is considered over two overlapping subdomains of different overlapping size $L_2 - L_1$, to conclude on the accuracy of the algorithm with the operator-splitting method. The considered subdomains are $\Omega_1 = [0, 60]$ and $\Omega_2 = [30, 150]$, as well as $\Omega_1 = [0, 100]$ and $\Omega_2 = [30, 150]$, see Figure 6.3. We apply the maximum norm, L_∞-norm, as an error norm for the error between the exact and numerical solution.

Decomposed Domains

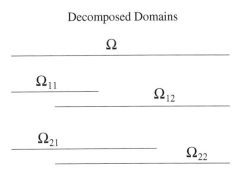

FIGURE 6.3: Overlapping situations for the numerical example.

The graphical output for the FOP method is presented in Figure 6.4. In our numerical computations, we refined the time- and space-steps systematically in order to visualize the accuracy and error reduction throughout the

TABLE 6.12: The L_∞-error in time and space for the convection-diffusion-reaction equation using FOP method.

time-step	err_{u,L_∞}	err_{u,L_∞}	err_{u,L_∞}
$\tau = 5$	0.001108	2.15813e-4	6.55262e-5
$\tau = 10$	0.00113	2.3942e-4	8.6641e-5
$\tau = 20$	0.001195	2.86514e-4	1.2868e-4
Spatial step	$\Delta x = 1$	$\Delta x = 0.5$	$\Delta x = 0.25$

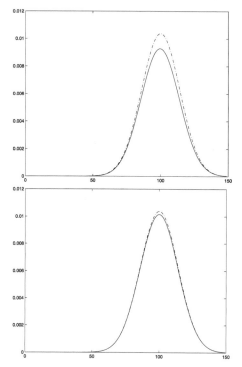

FIGURE 6.4: The results for the FOP method plotted as solid line in comparison with the exact solution plotted as dashed line, for $\Delta x = 1$ (left figure) and $\Delta x = 0.5$ (right figure).

simulation over the time interval for refined time- and space-steps. In Table 6.13 we observe that with the FOP method, the error reduced in first and second order with respect to time and space.

For the solution with the FOPSWR method, using FOP method as basic solver, the accuracy of the solution improved as the size of overlapping subdomains increased.

The results for the modified method are presented in Figure 6.5.

TABLE 6.13: The L_∞-error in time and space for the scalar convection-diffusion-reaction equation using FOPSWR method with two different sizes of overlapping: 30 and 70.

Time-step	$\mathrm{err}_{u,L_\infty}$	$\mathrm{err}_{u,L_\infty}$	$\mathrm{err}_{u,L_\infty}$	$\mathrm{err}_{u,L_\infty}$	$\mathrm{err}_{u,L_\infty}$	$\mathrm{err}_{u,L_\infty}$
$\tau = 5$	1.108e-3	1.08e-3	2.159e-4	2.158e-4	6.782e-5	6.552e-5
$\tau = 10$	1.138e-3	1.137e-3	2.397e-4	2.394e-4	8.681e-5	8.681e-5
$\tau = 20$	1.196e-3	1.195e-3	2.871e-4	2.865e-4	1.290e-4	1.286e-4
overlap.	30	70	30	70	30	70
Spatial step	$\Delta x = 1$		$\Delta x = 0.5$		$\Delta x = 0.25$	

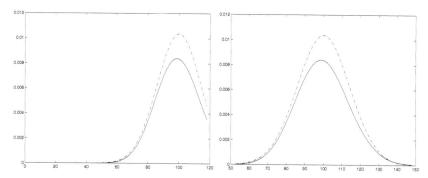

FIGURE 6.5: The results for FOPSWR method with overlapping size 30 plotted as solid line and the exact solution plotted as dashed line.

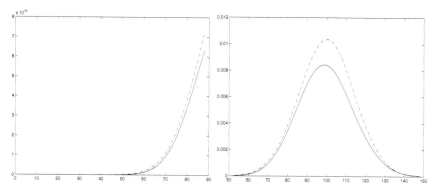

FIGURE 6.6: The results for FOPSWR method with overlapping size 70 plotted as solid line and the exact solution plotted as dashed line.

6.3.2 Second Example: System of Convection-Diffusion-Reaction Equations (Decoupled) Solved with Operator-Splitting Methods

We consider a more complicated example of a one-dimensional convection-diffusion-reaction equation:

$$R_i\,\partial_t u_i + v\,\partial_x u_i - \partial_x\,D\,\partial_x u_i = -R_i\lambda_i u_i, \tag{6.41}$$

$$u_i(x, t_0) = u^i_{\mathrm{exact}}(x, t_0),\ u_i(\pm\infty, t) = 0, \tag{6.42}$$

with the analytical solution given as (cf. [94])

$$u_{\text{exact}}^i(x,t) = \frac{u_{in}^i}{2R_i\sqrt{D\,\pi\,t/R_i}}\,\exp(-\frac{(x - v\,t/R_i)^2}{4\,D\,t/R_i})\,\exp((-\lambda_i\,t)),\quad(6.43)$$

for $i = 1$ and $i = 2$.

For the initial conditions we have the parameters $u_{in}^1 = 1.0$ and $u_{in}^2 = 1.0$ and the start time $t_0 = 100$. For boundary condition, we use the Dirichlet boundary condition with 0.0. We have the equation parameters $\lambda_1 = 1.0 \cdot 10^{-5}$, $\lambda_2 = 4.0 \cdot 10^{-5}$, $v = 0.001$, $D = 0.0001$, $R_1 = 2.0$, and $R_2 = 1.0$. The time interval is defined with $t \in [t_0, t_{\text{end}}]$ and $t_{\text{end}} = 10^5$.

The results for the classical method (operator-splitting method) are given in Table 6.14. In the next experiments, we consider the modified method.

TABLE 6.14: The L_∞-error in time and space for the system of decoupled convection-diffusion-reaction equations using FOP method.

Time-step	$\text{err}_{u_1,L_\infty}$	$\text{err}_{u_2,L_\infty}$	$\text{err}_{u_1,L_\infty}$	$\text{err}_{u_2,L_\infty}$	$\text{err}_{u_1,L_\infty}$	$\text{err}_{u_2,L_\infty}$
$\tau = 5$	4.461e-4	2.045e-4	3.466e-4	1.940e-4	9.110e-5	1.792e-4
$\tau = 10$	4.506e-4	2.055e-04	3.515e-4	1.948e-4	9.528e-5	1.794e-4
$\tau = 20$	4.594e-4	2.075e-4	3.611e-4	1.963e-4	1.036e-4	1.799e-4
Spatial step	$\Delta x = 1$		$\Delta x = 0.5$		$\Delta x = 0.25$	

For the overlapping part, we obtain the overlap sizes of 30 and 70 that is, $\Omega_1 = \{0 < x < 60\}$ and $\Omega_2 = \{30 < x < 150\}$, while for the other case we have $\Omega_1 = \{0 < x < 100\}$ and $\Omega_2 = \{30 < x < 150\}$.

The results are given in Table 6.15.

TABLE 6.15: The L_∞-error in time and space for the system of decoupled convection-diffusion-reaction equations using FOPSWR method with the overlapping size of 30.

Time-step	$\text{err}_{u_1,L_\infty}$	$\text{err}_{u_2,L_\infty}$	$\text{err}_{u_1,L_\infty}$	$\text{err}_{u_2,L_\infty}$	$\text{err}_{u_1,L_\infty}$	$\text{err}_{u_2,L_\infty}$
$\tau = 5$	4.461e-4	2.046e-4	3.466e-4	1.941e-4	9.110e-5	1.792e-4
$\tau = 10$	4.506e-4	2.056e-04	3.515e-4	1.948e-4	9.528e-5	1.795e-4
$\tau = 20$	4.594e-4	2.076e-4	3.611e-4	1.964e-4	1.036e-4	1.800e-4
Spatial step	$\Delta x = 1$		$\Delta x = 0.5$		$\Delta x = 0.25$	
overlap.	30					

The next result is given with the overlapping size of 70, see Table 6.16. By comparing the results, we see that we can improve the computations with the

TABLE 6.16: The L_∞-error in time and space for the system of decoupled convection-diffusion-reaction equations using FOPSWR method with the overlapping size of 70.

Time-step	$\text{err}_{u_1,L_\infty}$	$\text{err}_{u_2,L_\infty}$	$\text{err}_{u_1,L_\infty}$	$\text{err}_{u_2,L_\infty}$	err_{u_1}	err_{u_2}
$\tau = 5$	4.461e-4	2.046e-4	3.466e-4	1.941e-4	9.110e-5	1.792e-4
$\tau = 10$	4.506e-4	2.055e-04	3.515e-4	1.948e-4	9.528e-5	1.794e-4
$\tau = 20$	4.594e-4	2.075e-4	3.611e-4	1.963e-4	1.036e-4	1.800e-4
Spatial step	$\Delta x = 1$		$\Delta x = 0.5$		$\Delta x = 0.25$	
overlap.	70					

modified method and obtain higher-order results. Because of the decoupling, each equation could be computed separately. For the first component, we derive improved results because of the smaller reaction in the equation.

6.3.3 Third Example: System of Convection-Diffusion-Reaction Equations (Coupled) Solved with Operator-Splitting Methods

We deal with the more complicated example of a one-dimensional convection-diffusion-reaction equation:

$$R_1 \partial_t u_1 + v \partial_x u_1 - \partial_x D \partial_x u_1 = -R_1 \lambda_1 u_1, \tag{6.44}$$

$$R_2 \partial_t u_2 + v \partial_x u_2 - \partial_x D \partial_x u_2 R_1 \lambda_1 u_1 - R_2 \lambda_2 u_2, \tag{6.45}$$

$$u_1(x,t_0) = u_{\text{exact}}^1(x,t_0) , \; u_1(\pm\infty,t) = 0,$$

$$u_2(x,t_0) = u_{\text{exact}}^2(x,t_0) , \; u_2(\pm\infty,t) = 0,$$

The analytical solution is given as (cf. [94])

$$u_{\text{exact}}^1(x,t) = \frac{u_{in}^1}{2R_1\sqrt{D\,\pi\,t/R_1}} \, \exp(-\frac{(x - v\,t/R_1)^2}{4\,D\,t/R_1}) \, \exp((-\lambda_1\,t)),$$

$$u_{\text{exact}}^2(x,t) = \frac{u_{in}^2}{2\,R_2\,\sqrt{D\,\pi\,t/R_2}} \, \exp(-\frac{(x - v\,t/R_2)^2}{4\,D\,t/R_2}) \, \exp((-\lambda_2\,t))$$

$$+ \frac{R_1\,\lambda_1\,u_{in}^1}{2\sqrt{D\,\pi}\,(R_1 - R_2)} \, \exp(\frac{xv}{2D}) \, \exp(\frac{-(R_1\,\lambda_1 - R_2\,\lambda_2)\,t}{(R_1 - R_2)})(W(\mathrm{v}_2) - W(\mathrm{v}_1)),$$

$$\mathrm{v}_1 = \sqrt{R_1\,\lambda_1 - \frac{(R_1\,\lambda_1 - R_2\,\lambda_2)}{R_1 - R_2}\,R_1 + v^2/(4D)} \, ,$$

$$\mathrm{v}_2 = \sqrt{R_2\,\lambda_2 - \frac{(R_1\,\lambda_1 - R_2\,\lambda_2)}{R_1 - R_2}\,R_2 + v^2/(4D)} \, ,$$

$$W(\mathrm{v}) = 0.5(\exp(-\frac{x\,v\,\mathrm{v}}{2\,D})\mathrm{erfc}(\frac{x - v\,\mathrm{v}\,t}{\sqrt{4\,Dt}}) + \exp(\frac{x\,v\,\mathrm{v}}{2\,D})\mathrm{erfc}(\frac{x + v\,\mathrm{v}\,t}{\sqrt{4\,Dt}})) \, ,$$

where erfc($\cdot$) is the known error function, see [1], and we have the following conditions: $R_1 > R_2$ and $\lambda_2 > \lambda_1$.

For the initial conditions we have the parameters $u_{in}^1 = 1.0$, $u_{in}^2 = 0.0$, and the start time $t_0 = 100$. For the boundary condition, we use the Dirichlet boundary condition with the value 0.0. We have the equation parameters $\lambda_1 = 1.0 \cdot 10^{-5}$, $\lambda_2 = 4.0 \cdot 10^{-5}$, $v = 0.001$, $D = 0.0001$, $R_1 = 2.0$, and $R_2 = 1.0$. The initial conditions are given as $u_{10} = 1.0$ and $u_{20} = 0.0$. The time interval is defined by $t \in [t_0, t_{end}]$ and $t_{end} = 10^5$.

In the following tables, we compare the classical with the modified method, and we test different reaction parameters and different splitting orders.

The results for the classical method (operator-splitting method) are given in Table 6.17.

TABLE 6.17: The L_∞-error in time and space for the system of convection-diffusion-reaction equations using FOP method, with $\lambda_1 = 2.0 \cdot 10^{-5}$, $\lambda_2 = 4.0 \cdot 10^{-5}$.

Time-step	$\mathrm{err}_{u_1,L_\infty}$	$\mathrm{err}_{u_2,L_\infty}$	$\mathrm{err}_{u_1,L_\infty}$	$\mathrm{err}_{u_2 L_\infty}$	$\mathrm{err}_{u_1,L_\infty}$	$\mathrm{err}_{u_2,L_\infty}$
$\tau = 5$	4.461e-4	2.403e-3	3.466-4	2.452e-3	9.110e-5	2.702e-3
$\tau = 10$	4.506e-4	2.39e-3	3.515e-4	2.447e-3	9.528e-5	2.697e-3
$\tau = 20$	4.594e-4	2.8e-3	3.611e-4	2.438e-3	1.036e-4	2.689e-3
Spatial step	$\Delta x = 1$		$\Delta x = 0.5$		$\Delta x = 0.25$	

The results for the modified method (operator-splitting method and domain decomposition method) are given in Table 6.18.

TABLE 6.18: The L_∞-error in time and space for the system of convection-diffusion-reaction equations using FOPSWR method, with $\lambda_1 = 2 \cdot 10^{-5}$, $\lambda_2 = 4 \cdot 10^{-5}$.

Time-step	$\mathrm{err}_{u_1,L_\infty}$	$\mathrm{err}_{u_2,L_\infty}$	$\mathrm{err}_{u_1,L_\infty}$	$\mathrm{err}_{u_2,L_\infty}$	$\mathrm{err}_{u_1,L_\infty}$	$\mathrm{err}_{u_2,L_\infty}$
$\tau = 5$	4.461e-4	2.403e-3	3.466e-4	2.452e-3	9.110e-5	2.702e-3
$\tau = 10$	4.506e-4	2.398e-3	3.515e-4	2.447e-3	9.528e-5	2.697e-3
$\tau = 20$	4.594e-4	2.388e-3	3.611e-4	2.438e-3	1.036e-4	2.689e-3
Spatial step	$\Delta x = 1$		$\Delta x = 0.5$		$\Delta x = 0.25$	
overlap.	70					

We can compare the results and improve the computations using modified method to obtain higher-order results.

In Figure 6.7, the results for solutions with different time-steps are presented. We run a second experiment with modified reaction parameters to

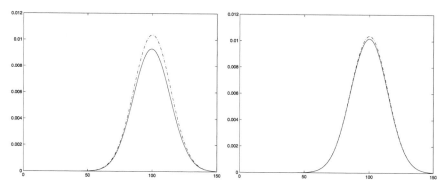

FIGURE 6.7: The results for FOP method plotted as solid line for the different steps $\Delta x = 1$ (left figure) and $\Delta x = 0.5$ (right figure) and the exact solution plotted as dashed line.

observe the influence of the first and the second components.

We use the classical second-order splitting method in the first computation; the results are shown in Table 6.19. In the second computation, we use the

TABLE 6.19: The L_∞-error in time and space for the system of convection-diffusion-reaction equations using SOP method, with $\lambda_1 = 1.0 \cdot 10^{-9}$, $\lambda_2 = 4.0 \cdot 10^{-5}$.

Time-step	$\mathrm{err}_{u_1,L_\infty}$	$\mathrm{err}_{u_2,L_\infty}$	$\mathrm{err}_{u_1,L_\infty}$	$\mathrm{err}_{u_2,L_\infty}$	$\mathrm{err}_{u_1,L_\infty}$	$\mathrm{err}_{u_2,L_\infty}$
$\tau = 5$	3.297e-3	6.058e-7	2.562e-3	6.192e-7	6.753e-4	6.820e-7
$\tau = 10$	3.30e-3	6.044e-7	2.599e-3	6.179e-7	7.083e-4	6.808e-7
$\tau = 20$	3.396e-3	6.018e-7	2.673e-3	6.152e-7	7.746e-4	6.784e-7
Spatial step	$\Delta x = 2$		$\Delta x = 1$		$\Delta x = 0.5$	

modified method and obtain the results shown in Table 6.20. In the Tables 6.19 and 6.20 we see higher-order results in space for the first component. The first component has an important influence on the second component, and we find that decreasing the error of the second component decreases the error of the first component. The results for the modified method are shown in Figure 6.18. The very strong coupling of the equation causes the methods to be of second order, because the error reduces and the computations are more effective.

TABLE 6.20:   The L_∞-error in time and space for the system of convection-diffusion-reaction equations using SOPSWR method with $\lambda_1 = 1.0 \cdot 10^{-9}, \quad \lambda_2 = 4 \cdot 10^{-5}$.

Time-step	$\mathrm{err}_{u_1,L_\infty}$	$\mathrm{err}_{u_2,L_\infty}$	$\mathrm{err}_{u_1,L_\infty}$	$\mathrm{err}_{u_2,L_\infty}$	$\mathrm{err}_{u_1,L_\infty}$	$\mathrm{err}_{u_2,L_\infty}$
$\tau = 5$	3.297e-3	6.058e-7	2.545e-3	6.192e-7	6.753e-4	6.820e-7
$\tau = 10$	3.314e-3	6.044e-7	2.599e-3	6.179e-7	7.083e-4	6.808e-7
$\tau = 20$	3.380e-3	6.018e-7	2.673e-3	6.152e-7	7.746e-4	6.784e-7
Spatial step	$\Delta x = 2$		$\Delta x = 1$		$\Delta x = 0.5$	
overlap.	70					

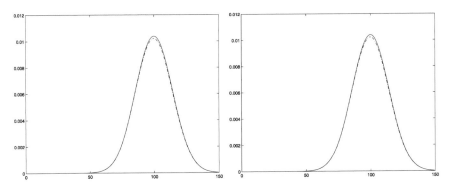

FIGURE 6.8:   The results for SOP method plotted as the solid line for different time-steps $\tau = 10$ (left figure) as well as $\tau = 20$ (right figure) and the exact solution is plotted with the dashed line.

REMARK 6.7 The efficient results are taken with the spatial and time decomposition methods. The first-order splitting methods in time are sufficient to reach the efficient results because of the higher spatial error. Second-order splitting in time and more waveform-relaxation steps in space can obtain finer spatial discretization and reach higher-order results. □

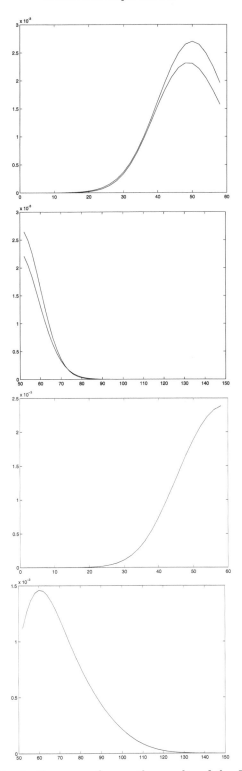

FIGURE 6.9:   In the upper figures the results of the first component
performed with SOPSWR method and plotted with the lower solid line and
the analytical solution plotted with the upper solid line are shown, where in

6.4 Benchmark Problems for the Temporal Decomposition Methods for Hyperbolic Equations

In this section, we will present the benchmark problems for the time decomposition methods for hyperbolic equations.

We are interested in the spatial dependent wave equation:

$$
\begin{aligned}
\frac{\partial^2 u}{\partial t^2} &= D_1(x)\frac{\partial^2 u}{\partial x_1^2} + \ldots + D_d(x)\frac{\partial^2 u}{\partial x_d^2}, \ t \in [0,T], x \in \Omega, \\
u(x,0) &= u_0(x), \ x \in \Omega, \\
\partial_t u(x,0) &= u_1(x), \ x \in \Omega, \\
u(x,t) &= u_2(x,t), \ t \in [0,T], x \in \partial\Omega_D, \\
\partial_n u(x,t) &= 0, \ t \in [0,T], x \in \partial\Omega_N,
\end{aligned}
\tag{6.46}
$$

where $x = (x_1,\ldots,x_d)^T \in \Omega$, the initial functions are given as $u_0, u_1 : \Omega \to \mathbb{R}^+$, and the function for the Dirichlet boundary is $u_2 : \Omega \times [0,T] \to \mathbb{R}^+$. The domain $\Omega \subset \mathbb{R}^d$ is Lipschitz continuous and convex, and the time interval is $[0,T] \subset \mathbb{R}^+$. The boundary is given as $\partial\Omega = \partial\Omega_D \cup \partial\Omega_N$.

For constant wave parameters (i.e., $D_1,\ldots,D_d \in \mathbb{R}^+$) we can derive the analytical solution given as

$$
u(x_1,\ldots,x_d,t) = \sin(\frac{1}{\sqrt{D_1}}\pi x_1) \cdot \ldots \cdot \sin(\frac{1}{\sqrt{D_d}}\pi x_d) \cdot \cos(\sqrt{d}\pi t). \tag{6.47}
$$

We compute a reference solution for the nonconstant wave parameters with sufficiently fine grids and small time-steps, see [178].

We now discuss the noniterative and iterative splitting methods for two-dimensional and three-dimensional wave equations.

6.4.1 Numerical Examples of the Elastic Wave Propagation with Noniterative Splitting Methods

In the following we present the simulations of test examples for the wave equation with noniterative splitting methods.

The noniterative splitting method in the time discretized form is given as

$$
\begin{aligned}
&\tilde{u} - 2u^n + u^{n-1} \\
&= \tau_n^2 D_1 \left(\eta \frac{\partial^2 \tilde{u}}{\partial x^2} + (1 - 2\eta) \frac{\partial^2 u^n}{\partial x^2} + \eta \frac{\partial^2 u^{n-1}}{\partial x^2} \right) \\
&\quad + \tau_n^2 D_2 \frac{\partial^2 u^n}{\partial y^2},
\end{aligned}
\tag{6.48}
$$

$$
\begin{aligned}
&u^{n+1} - 2u^n + u^{n-1} \\
&= \tau_n^2 D_1 \left(\eta \frac{\partial^2 \tilde{u}}{\partial x^2} + (1 - 2\eta) \frac{\partial^2 u^n}{\partial x^2} + \eta \frac{\partial^2 u^{n-1}}{\partial x^2} \right) \\
&\quad + \tau_n^2 D_2 \left(\eta \frac{\partial^2 u^{n+1}}{\partial y^2} + (1 - 2\eta) \frac{\partial^2 u^n}{\partial y^2} + \eta \frac{\partial^2 u^{n-1}}{\partial y^2} \right),
\end{aligned}
\tag{6.49}
$$

where $\tau_n = t^{n+1} - t^n$ is the time-step and the time discretization is of second order.

6.4.1.1 First Test Example: Wave Equation with Constant Coefficients

The application of the different boundary conditions is important for the splitting method used for the test example.

Dirichlet Boundary Condition
Our example is two-dimensional, where we can derive an analytical solution:

$$
\partial_{tt} u = D_1 \partial_{xx} u + D_2 \partial_{yy} u, \text{ in } \Omega \times [0, T], \tag{6.50}
$$
$$
u(x, y, 0) = u_{\text{exact}}(x, y, 0), \text{ on } \Omega,
$$
$$
\partial_t u(x, y, 0) = 0, \text{ on } \Omega,
$$
$$
u(x, y, t) = u_{\text{exact}}(x, y, t), \text{ on } \partial\Omega \times (0, T), \tag{6.51}
$$

where $\Omega = [0, 1] \times [0, 1]$, $D_1 = 1$, $D_2 = 0.5$.

The analytical solution is given as

$$
u_{\text{exact}}(x, y, t) = \sin(\frac{1}{\sqrt{D_1}} \pi x) \sin(\frac{1}{\sqrt{D_2}} \pi y) \cos(\sqrt{2}\, \pi t). \tag{6.52}
$$

The discretization is given with the implicit time discretization and the finite difference method for the space discretization.

Thus, we have for the second-order discretization in space

$$
\begin{aligned}
Au(t) &= D_1 \partial_{xx} u(t) \\
&\approx D_1 \frac{u(x + \Delta x, y, t) - 2u(x, y, t) + u(x - \Delta x, y, t)}{\Delta x^2},
\end{aligned}
\tag{6.53}
$$

$$
\begin{aligned}
Bu(t) &= D_2 \partial_{yy} u(t) \\
&\approx D_2 \frac{u(x, y + \Delta y, t) - 2u(x, y, t) + u(x, y - \Delta y, t)}{\Delta y^2},
\end{aligned}
\tag{6.54}
$$

and the second-order time discretization

$$\partial_{tt} u(t^n) \approx \frac{u(t^n + \tau_n) - 2u(t^n) + u(t^n - \tau_n)}{\tau_n^2}. \tag{6.55}$$

The implicit discretization is

$$u(t^{n+1}) - 2u(t^n) + u(t^{n-1}) \tag{6.56}$$
$$= \tau_n^2 (A + B)(\eta u(t^{n+1}) + (1 - 2\eta)u(t^n) + \eta u(t^{n-1})).$$

For the approximation error, we choose the L_1-norm, also the L_2- and L_∞-norm are possible.

The error in the L_1-norm is given as

$$err_{L_1} := \sum_{i,j=1,\ldots,m} V_{i,j} \, |u(x_i, y_j, t^n) - u_{\text{exact}}(x_i, y_j, t^n)|, \tag{6.57}$$

where $u(x_i, y_j, t^n)$ is the numerical and $u_{\text{exact}}(x_i, y_j, t^n)$ is the analytical solution. $V_{i,j} = \Delta x \, \Delta y$ is the mass, with Δx and Δy the equidistant grid steps in the x- and y-dimensions. m is the number of the grid nodes in the x- and y-dimensions.

In our test example we choose $D_1 = D_2 = 1$ (the Dirichlet boundary) and set our model domain to be a rectangle $\Omega = [0, 1] \times [0, 1]$.

We discretize with $\Delta x = 1/16$, $\Delta y = 1/16$, and $\tau_n = 1/32$, and choose our parameter η between $0 \le \eta \le 0.5$.

The experimental results of the finite differences, the classical operator-splitting method, and the LOD method are shown in Tables 6.21, 6.22, and 6.23 and Figure 6.10.

TABLE 6.21: Numerical results for finite differences method (second-order finite differences in time and space) and a Dirichlet boundary.

η	err_{L_1}	u_{exact}	u_{num}
0.0	0.0014	-0.2663	-0.2697
0.1	0.0030	-0.2663	-0.2738
0.3	0.0063	-0.2663	-0.2820
0.5	0.0096	-0.2663	-0.2901

REMARK 6.8 In these experiments, we compare the three different methods based on the ADI method. The splitting methods have the same

TABLE 6.22: Numerical results for classical operator-splitting method (second-order ADI method) and a Dirichlet boundary.

η	err_{L_1}	u_{exact}	u_{num}
0.0	0.0014	−0.2663	−0.2697
0.1	0.0030	−0.2663	−0.2738
0.3	0.0063	−0.2663	−0.2820
0.5	0.0096	−0.2663	−0.2901

TABLE 6.23: Numerical results for LOD method and a Dirichlet boundary.

η	err_{L_1}	err_{L_∞}
0.0	0.0014	0.0034
0.1	0.0031	0.0077
0.3	0.0065	0.0161
0.5	0.0099	0.0245

error reduction quality, while the most accurate method is the LOD method, which is a fourth-order method. The splitting methods are more effective methods. ▯

Neumann Boundary Condition

Our next example is two-dimensional, as we saw in the first example, but with respect to the Neumann boundary conditions:

$$\partial_{tt}u = D_1\partial_{xx}u + D_2\partial_{yy}u, \text{ in } \Omega \times (0,T), \tag{6.58}$$

$$u(x,y,0) = u_{\text{exact}}(x,y,0), \text{ on } \Omega,$$

$$\partial_t u(x,y,0) = 0, \text{ on } \Omega,$$

$$\frac{\partial u(x,y,t)}{\partial n} = \frac{\partial u_{\text{exact}}(x,y,t)}{\partial n}, \text{ on } \partial\Omega \times (0,T), \tag{6.59}$$

where $\Omega = [0,1] \times [0,1]$, $D_1 = 1$, $D_2 = 0.5$, and we have an equidistant time-step τ_n.

The analytical solution is given as

$$u_{\text{exact}}(x,y,t) = \sin(\frac{1}{\sqrt{D_1}}\pi x)\sin(\frac{1}{\sqrt{D_2}}\pi y)\cos(\sqrt{2}\,\pi t). \tag{6.60}$$

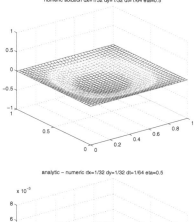

FIGURE 6.10: Numerical resolution of the wave equation: the numerical approximation (left figure) and error functions (right figure) for the Dirichlet boundary ($\Delta x = \Delta y = 1/32$, $\tau_n = 1/64$, $D_1 = 1$, $D_2 = 1$, coupled version).

The experimental results of the finite differences method and classical operator-splitting method are shown in Tables 6.24, 6.25 and Figure 6.11.

REMARK 6.9 The Neumann boundary conditions are more delicate because of the underlying boundary splitting methods. ⬚

6.4.1.2 Second Test Example: Spatial Dependent Test Example

In this experiment, we apply our method to the spatial dependent problem, given by

$$\partial_{tt} u = D_1(x, y)\partial_{xx} u + D_2(x, y)\partial_{yy} u, \text{ in } \Omega \times (0, T), \quad (6.61)$$
$$u(x, y, 0) = u_0(x, y), \text{ in } \Omega,$$
$$\partial_t u(x, y, 0) = u_1(x, y), \text{ in } \Omega,$$
$$u(x, y, t) = u_2(x, y, t), \text{ on } \partial\Omega \times (0, T),$$

where $D_1(x, y) = 0.1x + 0.01y + 0.01$, $D_2(x, y) = 0.01x + 0.1y + 0.1$.

TABLE 6.24: Numerical results for finite differences method (finite differences in time and space of second order) and a Neumann boundary.

η	err_{L_1}	u_{exact}	u_{num}
0.0	0.0014	−0.2663	−0.2697
0.1	0.0030	−0.2663	−0.2738
0.3	0.0063	−0.2663	−0.2820
0.5	0.0096	−0.2663	−0.2901
0.7	0.0128	−0.2663	−0.2981
0.9	0.0160	−0.2663	−0.3060
1.0	0.0176	−0.2663	−0.3100

TABLE 6.25: Numerical results for classical operator-splitting method (ADI method of second order) and a Neumann boundary.

η	err_{L_1}	u_{exact}	u_{num}
0.0	0.0014	−0.2663	−0.2697
0.1	0.0030	−0.2663	−0.2738
0.3	0.0063	−0.2663	−0.2820
0.5	0.0096	−0.2663	−0.2901

To compare the numerical results, we cannot use an analytical solution, which is why we compute a reference solution in an initial prestep. The reference solution is performed with the finite difference scheme, with fine time- and space-steps.

When choosing the time-steps, it is important to consider the CFL condition, which in this case is based on the spatial coefficients.

REMARK 6.10 We have assumed the following CFL condition:

$$\tau_n < 0.5 \frac{\min(\Delta x, \Delta y)}{\max_{x,y \in \Omega}(D_1(x,y), D_2(x,y))}, \tag{6.62}$$

where Δx and Δy are equidistant spatial steps for the x- and y-dimensions. $\square$

For the test example, we define our model domain as a rectangle $\Omega = [0,1] \times [0,1]$.

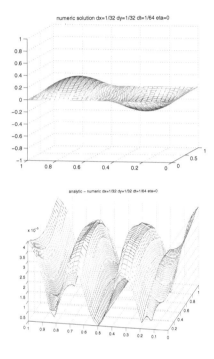

FIGURE 6.11: Numerical resolution of the wave equation: the numerical approximation (left figure) and error functions (right figure) for the Neumann boundary (right) ($\Delta x = \Delta y = 1/32$, $\tau_n = 1/64$, $D_1 = 1$, $D_2 = 0.5$, coupled version).

The reference solution is obtained by executing the finite differences method and setting $\Delta x = 1/256$, $\Delta y = 1/256$ and a time-step $\tau_n = 1/256 < 0.390625$.

The model domain is given by a rectangle with $\Delta x = 1/16$ and $\Delta y = 1/32$. The time-steps are given by $\tau_n = 1/16$ and $0 \leq \eta \leq 0.5$.

The numerical results for the Dirichlet boundary conditions are given in Tables 6.26, 6.27, 6.28 and Figure 6.12.

The results show the second-order accuracy and the similar results of the nonsplitting method (see Table 6.26) and classical splitting method (see Table 6.27). Therefore, the splitting method did not influence the numerical results, as a preserving second-order method. The results can be improved by the LOD method (see Table 6.28), while it reaches fourth-order accurate results with the parameter $\eta = \frac{1}{12}$.

The numerical results for the Neumann boundary conditions are given in Tables 6.29, 6.30 and Figure 6.13. We obtain the same results as shown in the Dirichlet case. The second-order accuracy of the classical splitting method (see Table 6.30) did not influence the second-order numerical results. The

TABLE 6.26: Numerical results for finite differences method with spatial dependent parameters and a Dirichlet boundary (error of the reference solution).

η	err_{L_1}	u_{exact}	u_{num}
0.0	0.0032	−0.7251	−0.7154
0.1	0.0034	−0.7251	−0.7149
0.3	0.0037	−0.7251	−0.7139
0.5	0.0040	−0.7251	−0.7129

TABLE 6.27: Numerical results for classical operator-splitting method with spatial dependent parameters and a Dirichlet boundary (error of the reference solution).

η	err_{L_1}	u_{exact}	u_{num}
0.0	0.0032	−0.7251	−0.7154
0.1	0.0034	−0.7251	−0.7149
0.3	0.0037	−0.7251	−0.7139
0.5	0.0040	−0.7251	−0.7129

TABLE 6.28: Numerical results for LOD method with spatial dependent parameters and a Dirichlet boundary (error of the reference solution).

η	err_{L_1}	u_{exact}	u_{num}
0.00	0.0032	−0.7251	−0.7154
0.1	0.7809e-003	−0.7251	−0.7226
0.122	0.6793e-003	−0.7251	−0.7242
0.3	0.0047	−0.7251	−0.7369
0.5	0.0100	−0.7251	−0.7512

same improvement for the Neumann case to reach higher order can be done with the LOD method, see [104].

REMARK 6.11 In our experiments, we have analyzed both the classical operator splitting and the LOD method, and showed that the LOD

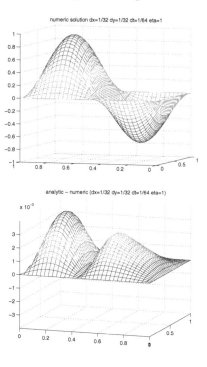

FIGURE 6.12: Dirichlet boundary condition: numerical solution (right figure) and error function (left figure) for the spatial dependent test example.

TABLE 6.29: Numerical results for finite differences method with spatial dependent parameters and Neumann boundary (error of the reference solution).

η	err_{L_1}	u_{exact}	u_{num}
0.0	0.0180	-0.7484	-0.7545
0.1	0.0182	-0.7484	-0.7532
0.3	0.0185	-0.7484	-0.7504
0.5	0.0190	-0.7484	-0.7477

method yields more accurate values. ⬜

REMARK 6.12 We have presented different time splitting methods for the spatial dependent case of the wave equation. The contributions of this chapter concern the boundary splitting and the stiff operator treatment. For

TABLE 6.30: Numerical results for classical operator-splitting method with spatial dependent parameters and Neumann boundary (error of the reference solution).

η	err_{L_1}	u_{exact}	u_{num}
0.0	0.0180	-0.7484	-0.7545
0.1	0.0182	-0.7484	-0.7532
0.3	0.0185	-0.7484	-0.7504
0.5	0.0190	-0.7484	-0.7477

the boundary splitting method, we discussed the theoretical background, and the experiments showed that the method is also stable for the stiff case. We presented stable results even for the spatial dependent wave equation. The computational process benefits from decoupling the stiff and nonstiff operators into different equations, due to the different scales of the operators. The LOD method as a fourth-order method has the advantage of higher accuracy and can be used for such decoupling. In a future work, we will discuss the algorithms based on the eigenmodes of the operators for more flexibility in decoupling problems. ▯

6.4.2 Numerical Examples of the Elastic Wave Propagation with Iterative Splitting Methods

We now apply the two- and three-dimensional iterative operator-splitting methods to our wave equations. The time discretization is of second order, and our splitting method is given for the two-dimensional equations as

$$
\begin{aligned}
u^i(t^{n+1}) &- 2u^n + u^{n-1} \\
&= \tau_n^2 D_1 \left(\eta \frac{\partial^2 u^i(t^{n+1})}{\partial x^2} + (1-2\eta) \frac{\partial^2 u^n}{\partial x^2} + \eta \frac{\partial^2 u^{n-1}}{\partial x^2} \right) \quad (6.63) \\
&+ \tau_n^2 D_2 \left(\eta \frac{\partial^2 u^{i-1}(t^{n+1})}{\partial y^2} + (1-2\eta) \frac{\partial^2 u^n}{\partial y^2} + \eta \frac{\partial^2 u^{n-1}}{\partial y^2} \right),
\end{aligned}
$$

$$
\begin{aligned}
u^{i+1}(t^{n+1}) &- 2u^n + u^{n-1} \\
&= \tau_n^2 D_1 \left(\eta \frac{\partial^2 u^i(t^{n+1})}{\partial x^2} + (1-2\eta) \frac{\partial^2 u^n}{\partial x^2} + \eta \frac{\partial^2 u^{n-1}}{\partial x^2} \right) \quad (6.64) \\
&+ \tau_n^2 D_2 \left(\eta \frac{\partial^2 u^{i+1}(t^{n+1})}{\partial y^2} + (1-2\eta) \frac{\partial^2 u^n}{\partial y^2} + \eta \frac{\partial^2 u^{n-1}}{\partial y^2} \right),
\end{aligned}
$$

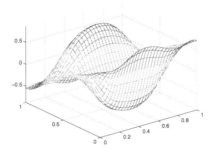

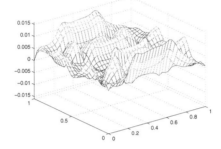

FIGURE 6.13: Neumann boundary condition: numerical solution (left figure) and error function (right figure) for the spatial dependent test example.

where $i = 1, 3, \ldots 2m + 1$, and the previous solutions are given as u^{n-1} and u^n, the starting solution $u^{i-1}(t^{n+1})$ is given as

$$u^{i-1}(t^{n+1}) - 2u^n + u^{n-1} = \tau_n^2 \left(D_1 \frac{\partial^2 u^n}{\partial x^2} + D_2 \frac{\partial^2 u^n}{\partial y^2} \right).$$

The error is calculated by

$$\mathrm{err}_{L_1} := \sum_{i=1}^{p} \Delta x_1 \cdot \ldots \cdot \Delta x_d \, |u(x_{1,i}, \ldots, x_{d,i}, t^n) - u_{\mathrm{exact}}(x_{1,i}, \ldots, x_{d,i}, t^n)|,$$

$$(6.65)$$

where p is the number of grid points. u_{exact} is the analytical solution, and u is the numerical solution; $\Delta x_1, \ldots, \Delta x_d$ are the equidistant spatial steps of each dimension.

In the following we present the simulations of test examples for the wave equation with iterative splitting methods.

6.4.2.1 First Test Example: Wave Equation with Constant Coefficients in Two Dimensions

In the first example, we compare the classical splitting method (ADI method) and the iterative splitting methods for a nonstiff and a stiff case.
We compare the nonstiff case for the operators A and B.

We begin with $D_1 = 1$, $D_2 = \frac{1}{4}$, $\Delta x = \frac{1}{16}$, $\Delta y = \frac{1}{32}$, two iterations per time-step, and $\Omega = [0,1]^2$, $t \in [0, 3 \cdot (1/\sqrt{2})]$.

We discretize with the temporal and spatial derivatives using second-order finite difference methods.

The results are given in Table 6.31 for the classical method (ADI method) and in Table 6.32 for the iterative operator-splitting method.

TABLE 6.31: Classical operator-splitting method; $D_1 = 1$, $D_2 = \frac{1}{4}$, $\Delta x = \frac{1}{16}$, $\Delta y = \frac{1}{32}$, and $\Omega = [0,1]^2$, $t \in [0, 3 \cdot (1/\sqrt{2})]$.

η	err_{L_1}	err_{L_1}	err_{L_1}
tsteps	44	45	46
0	0.0235	2.3944e-004	4.8803e-007
0.0100	2.0405e-004	5.3070e-007	7.9358e-008
0.0800	4.1689e-005	4.1885e-005	4.2068e-005
0.0833	4.5958e-005	4.5970e-005	4.5982e-005
0.1000	7.0404e-005	6.9235e-005	6.8149e-005
0.2000	3.2520e-004	3.0765e-004	2.9168e-004

TABLE 6.32: Iterative operator-splitting method; $D_1 = 1$, $D_2 = \frac{1}{4}$, $\Delta x = \frac{1}{16}$, $\Delta y = \frac{1}{32}$, two iterations per time-step, starting solution with an explicit time-step, and $\Omega = [0,1]^2$, $t \in [0, 3 \cdot (1/\sqrt{2})]$.

η	err_{L_1}	err_{L_1}	err_{L_1}
tsteps	44	45	46
0	0.0230	2.3977e-004	4.9793e-007
0.0100	1.4329e-004	1.2758e-006	7.9315e-008
0.0800	4.1602e-005	4.1805e-005	4.1994e-005
0.0833	4.5858e-005	4.5879e-005	4.5898e-005
0.1000	7.0228e-005	6.9074e-005	6.8003e-005
0.2000	3.2375e-004	3.0636e-004	2.9052e-004

In the next computations we compare the stiff case for the operators A and B.

We deal with $D_1 = \frac{1}{1}$, $D_2 = \frac{1}{100}$, $\Delta x = \frac{1}{16}$, $\Delta y = \frac{1}{32}$, and $\Omega = [0,1]^2 \times [0, 3 \cdot (1/\sqrt{2})]$. We discretize with the time and space derivatives with second-order finite difference methods.

The results are given in Table 6.33 for the classical method (ADI method) and in Table 6.34 for the iterative operator-splitting method.

TABLE 6.33: Classical operator-splitting method; $D_1 = \frac{1}{100}$, $D_2 = \frac{1}{100}$, $\Delta x = \frac{1}{16}$, $\Delta y = \frac{1}{32}$, and $\Omega = [0,1]^2$, $t \in [0, 3 \cdot (1/\sqrt{2})]$.

η	err_{L_1}	err_{L_1}	err_{L_1}	err_{L_1}	err_{L_1}	err_{L_1}
tsteps	3	5	6	7	15	100
0.0	41.0072	0.1205	9.2092e-004	0.0122	0.1131	0.1475
0.01	29.0629	0.0626	5.5714e-004	0.0212	0.1171	0.1476
0.05	6.9535	0.0100	0.0491	0.0764	0.1336	0.1479
0.08	0.9550	0.0984	0.1226	0.1331	0.1465	0.1482
0.09	0.1081	0.1409	0.1515	0.1539	0.1508	0.1483
0.1	0.3441	0.1869	0.1819	0.1755	0.1552	0.1484
0.2	0.7180	0.6259	0.4960	0.4052	0.2005	0.1494
0.3	0.2758	0.7949	0.7126	0.6027	0.2479	0.1504
0.4	0.0491	0.7443	0.7948	0.7316	0.2963	0.1513
0.5	0.0124	0.6005	0.7754	0.7905	0.3448	0.1523

TABLE 6.34: Iterative operator-splitting method; $D_1 = \frac{1}{100}$, $D_2 = \frac{1}{100}$, $\Delta x = \frac{1}{16}$, $\Delta y = \frac{1}{32}$, starting solution with an explicit time-step, and $\Omega = [0,1]^2$ $t \in [0, 3 \cdot (1/\sqrt{2})]$.

η	err_{L_1}	err_{L_1}	err_{L_1}	err_{L_1}	err_{L_1}	err_{L_1}
tsteps	3	5	6	7	15	100
# iter	4	8	10	12	28	198
0.0	11.5666	0.0732	6.1405e-004	0.0088	0.0982	0.1446
0.01	9.0126	0.0379	3.6395e-004	0.0151	0.1017	0.1447
0.05	3.3371	0.0047	0.0315	0.0540	0.1161	0.1450
0.08	1.4134	0.0520	0.0785	0.0935	0.1272	0.1453
0.09	0.9923	0.0756	0.0970	0.1081	0.1310	0.1454
0.1	0.6465	0.1019	0.1166	0.1231	0.1348	0.1455
0.2	0.7139	0.4086	0.3352	0.2877	0.1737	0.1465
0.3	0.7919	0.6541	0.5332	0.4466	0.2141	0.1474
0.4	0.6591	0.7963	0.6792	0.5787	0.2552	0.1484
0.5	0.5231	0.8573	0.7734	0.6791	0.2961	0.1493

REMARK 6.13 The iterative splitting method is in both cases more accurate, and the best results are given for the discretization parameter $\eta \approx 0.01$. Two iteration steps are sufficient to obtain the second-order results. For the stiff case, more iteration steps are important to obtain the same results as for the nonstiff case. ☐

6.4.2.2 Second Test Example: Wave Equation with Constant Coefficients in Three Dimensions

In the second example, we compare the iterative splitting methods for different starting solutions $u^{i-1}(t^{n+1})$.

We consider the explicit method first for the starting solution $u^{i-1,n+1}$, with

$$u^{i-1,n+1} - 2u^n + u^{n-1} = \tau_n^2 (D_1 \frac{\partial^2 u^n}{\partial x^2} + D_2 \frac{\partial^2 u^n}{\partial y^2}).$$

In the second method, we calculate $u^{i-1,n+1}$ using the classical operator-splitting method (ADI method), described in Section 6.4.1.

For our model equation, we apply the three-dimensional wave equation:

$$\frac{\partial^2 u}{\partial t^2} = D_1 \frac{\partial^2 u}{\partial x^2} + D_2 \frac{\partial^2 u}{\partial y^2} + D_3 \frac{\partial^2 u}{\partial z^2}, \tag{6.66}$$

where we have exact initial and boundary conditions given in the equation (6.47).

We use $D_1 = 1$, $D_2 = \frac{1}{10}$ and $D_3 = \frac{1}{100}$, $\Delta x = \Delta y = \Delta z = \frac{1}{8}$, three iterations per time-step, and $\Omega = [0,1]^3$, $t \in [0, 6 \cdot (1/\sqrt{3})]$. We discretize with the time and space derivatives with second-order finite difference methods.

The results are given in Table 6.35 for the classical method (ADI method) and in Table 6.36 for the iterative operator-splitting method.

TABLE 6.35: Iterative operator-splitting method; $D_1 = 1$, $D_2 = \frac{1}{10}$ and $D_3 = \frac{1}{100}$, $\Delta x = \Delta y = \Delta z = \frac{1}{8}$, three iterations per time-step, starting solution with an explicit time-step, and $\Omega = [0,1]^3$, $t \in [0, 6 \cdot (1/\sqrt{3})]$.

η	err_{L_1}	err_{L_1}	err_{L_1}	err_{L_1}	err_{L_1}
tsteps	8	9	10	11	12
0	169.4361	8.4256	0.5055	0.2807	0.3530
0.2000	0.0875	0.1315	0.1750	0.2232	0.2631
0.3000	0.1151	0.0431	0.0473	0.0745	0.1084
0.4000	0.3501	0.2055	0.0988	0.0438	0.0454
0.4500	0.4308	0.3002	0.1719	0.0844	0.0402
0.5000	0.4758	0.3792	0.2510	0.1402	0.0704

TABLE 6.36: Iterative operator-splitting method; $D_1 = 1$, $D_2 = \frac{1}{10}$ and $D_3 = \frac{1}{100}$, $\Delta x = \Delta y = \Delta z = \frac{1}{8}$, three iterations per time-step, starting solution with classical splitting method (ADI), and $\Omega = [0, 1]^3$, $t \in [0\ ,\ 5 \cdot (1/\sqrt{3})]$.

η	err$_{L_1}$	err$_{L_1}$	err$_{L_1}$	err$_{L_1}$	err$_{L_1}$
tsteps	6	7	8	9	10
0	79.0827	10.1338	0.2707	0.1981	0.2791
0.2000	0.1692	0.2658	0.3374	0.3861	0.4181
0.3000	0.0311	0.0589	0.1182	0.1983	0.2692
0.4000	0.1503	0.0400	0.0389	0.0793	0.1322
0.4500	0.2340	0.0832	0.0316	0.0502	0.0930
0.5000	0.3101	0.1401	0.0441	0.0349	0.0645

REMARK 6.14 The starting solution prestep combined with the ADI method is more accurate and gives the best results for the discretization parameter $\eta \approx 0.01$. In a three-dimensional case, one more iteration step is needed compared to the two-dimensional case. □

6.5 Real-Life Applications

In this section we deal with the real-life applications that were introduced in Chapter 1. We discuss the efficiency and accuracy of our proposed splitting methods for the multi-physics problems.

6.5.1 Waste Disposal: Transport and Reaction of Radioactive Contaminants

In the next subsections, we describe the two- and three-dimensional simulations of waste disposal.

6.5.1.1 Two-Dimensional Model of Waste Disposal

We calculate some waste disposal scenarios that help us to reach new conclusions about the waste disposals in salt domes.

We consider a model based on a salt dome with a layer of overlying rock, with a permanent source of groundwater flow that becomes contaminated with the radioactive waste. Based on our model, we calculate the transport and the reaction of these contaminants coupled with decay chains, see [76]. The simulation time is $10000[a]$, and we calculate the concentration of waste in the water that flows to the top of the overlying rock. With these dates we can determine if the waste disposal is safe. We present the two-dimensional test case with the dates of our last project partner GRS in Braunschweig (Germany), cf. [74] and [75].

We have a model domain with the size of $6000[m] \times 150[m]$ consisting of four different layers with different permeabilities, see [74]. Groundwater is pooled in the domain from the right to the left boundary. The groundwater flows faster through the permeable layer than through the impermeable layers. Therefore, the groundwater flows from the right boundary to the middle half of the domain. It flows through the permeable layer down to the bottom of the domain and pools at the top left of the domain to an outflow at the left boundary. The flow field with the velocity is calculated with the program package **D³F** (Distributed-Density-Driven-Flow software toolbox, see [72]), and is presented in Figure 6.14.

In the middle of the bottom of the domain, the contaminants flow as from a permanent source. With the stationary velocity field, the contaminants are computed with the software package **R³T** (Radioactive-Reaction-Retardation-Transport software toolbox, see [73]). The flow field transports the radioactive contaminants up to the top of the domain. The decay chain

FIGURE 6.14: Flow field for a two-dimensional calculation.

is presented with 26 components as follows,

$$Pu-244 \rightarrow Pu-240 \rightarrow U-236 \rightarrow Th-232 \rightarrow Ra-228$$
$$Cm-244 \rightarrow Pu-240$$
$$U-232$$
$$Pu-241 \rightarrow Am-241 \rightarrow Np-237 \rightarrow U-233 \rightarrow Th-229$$
$$Cm-246 \rightarrow Pu-242 \rightarrow U-238 \rightarrow U-234 \rightarrow Th-230 \rightarrow$$
$$Ra-226 \rightarrow Pb-210$$
$$Am-242 \rightarrow Pu-238 \rightarrow U-234$$
$$Am-243 \rightarrow Pu-239 \rightarrow U-235 \rightarrow Pa-231 \rightarrow Ac-227.$$

We present the important concentration in this decay chain. In Figure 6.15 the contaminant uranium isotope U-236 is presented after 100[a]. This movement of isotope is less retarded by adsorption than other nuclides and has a very long half-life period. The concentration of U-236 in the model domain is illustrated in Figure 6.15 at two different time points in the modeling. The diffusion process spreads the contaminant throughout the left side of the domain. The impermeable layer is also contaminated. After the time period of 10000[a], the contaminant is flown up to the top of the domain.

The calculations are performed on uniform grids. The convergence of these grids is confirmed with adaptive grid calculations, which confirmed the results of finer and smaller time-steps, cf. Table 6.37. The calculations begin with explicit methods until the character of the equation is more diffusive, at this point we switch to the implicit methods and use large time-steps. With this procedure, we can fulfill the mandatory maximum calculation time of 1 day.

6.5.1.2 Three-Dimensional Model of a Waste Disposal

In this example, we consider a three-dimensional model, because of the interest in the three-dimensional effects of contaminants in the flowing ground-

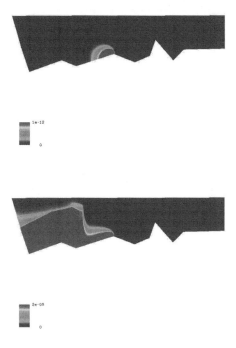

FIGURE 6.15: Concentration of U-236 at the time point $t = 100[a]$ and $t = 10000[a]$.

water. We simulate approximately $10000[a]$ and focus on the important contaminants flowing farthest with a high concentration. We assume an anisotropy domain of $6000[m] \times 2000[m] \times 1500[m]$ with different permeable layers. We have calculated 26 components as presented in the two-dimensional case. The parameters for the diffusion and dispersion tensor are given as $D = 1 \cdot 10^{-9}[m^2/s]$, $\alpha_L = 4.0\ [m]$, $\alpha_T = 0.4\ [m]$, $|v|_{max} = 6 \cdot 10^{-6}[m/s]$, $\rho = 2 \cdot 10^3$, $D_L = \alpha_L|v|$ and $D_T = \alpha_T|v|$, where the longitudinal dispersion length is ten times bigger than the transversal dispersion. The source is suited at the point $(4250.0, 2000.0, 1040.0)$, and the contaminants flow with a constant rate. The underlying velocity fields are calculated with $\mathbf{D}^3\mathbf{F}$, and we added two sinks at the surface with the coordinates $(2000, 2100, 2073)$ and $(2500, 2000, 2073)$.

We simulate the transport and the reaction of the contaminants with our software package $\mathbf{R}^3\mathbf{T}$. As a result, we can simulate the needed pumping rate of the sinks to pump out all of the contaminated groundwater. We present the velocity field in Figure 6.16. The groundwater flows from the right boundary to the middle of the domain. Due to the impermeable layers, the groundwa-

TABLE 6.37: Computing the two-dimensional case.

Processors	Refinement	Number of elements	Number of time-steps	Time for one time-step	total time
30	uniform	75000	3800	5 sec.	5.5 h.
64	adaptive	350000	3800	14 sec.	14.5 h.

ter flows downward and pools in the middle part of the domain. Because of the influence of the salt dome, salt wells up with the groundwater and becomes incorporated into the lower middle part of the domain curls. These parts are interesting for three-dimensional calculations, and because of the curls, the groundwater pools. We concentrate on the important component

FIGURE 6.16: Flow field for a three-dimensional calculation.

$U - 236$. This component is less retarded and is drawn into the sinks. In the top images of Figures 6.17, we show the initial concentration at time point $t = 100$ [a]. We present the data in vertical cut planes. In the next picture, we present a cut plane through the source term. In the bottom of Figures 6.17, the concentration is presented at the end time point $t = 10000$ [a]. The concentration has welled from the bottom up over the impermeable layer and into the sinks at the top of the domain.

At the beginning of the calculation, we use explicit discretization methods with respect to the convection-dominant case. After the initializing process, the contaminants are spread out with the diffusion process. We use the implicit methods with larger time-steps and could also calculate the mandatory time period with a higher-order discretization method.

The computations are performed on a Linux PC-Cluster with 1.6 GHz Athlon processors.

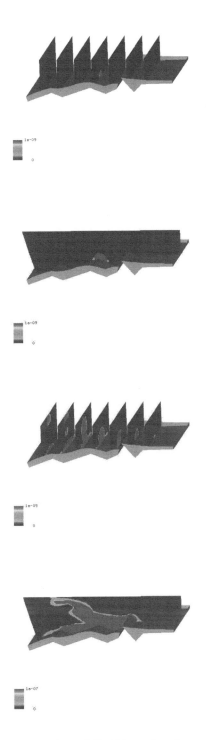

FIGURE 6.17: Concentration of U-236 at the time points $t = 100[a]$ and $t = 10000[a]$.

In Table 6.38, we report the results of the computations. We begin with convergence results on uniform refined meshes. We confirm these results with adaptive refined meshes and obtain the same results with smaller time-steps. We accomplished these calculations within the 1 day mandatory computational time limit.

TABLE 6.38:   Three-dimensional calculations of a realistic potential damage event.

Processors	Refinement	Number of elements	Number of time-steps	Time for one time-step	Total time
16	uniform	531264	3600	13.0 sec.	13.0 h
72	adaptive	580000	3600	18.5 sec.	18.5 h

REMARK 6.15 Here the performance is obvious, based on the decoupling of the equation into convection-reaction and diffusion parts. We could accelerate the solver process with the most adapted discretization and solver methods. The resources are saved with the direct ODE solvers for the reaction part and the possibility of large time-steps for the implicit discretized diffusion part. For such problems, a decoupling makes sense with respect to parallelizing the underlying solver methods.   ☐

6.5.2   Sublimation of Crystal Growth

In this next section, we focus on decoupling the complicated sublimation process in simpler processes. We discretize and solve the simpler equations by more accurate methods with embedded analytical solutions, see [83] and [84].

With the test examples, we verify the decomposition methods for the convection-reaction and heat equation. The models are based on the conduction of heat in solids, see [38] and a simulation of gas mixture, see [158]. Based on these results, we present a real-life application. We discuss the heat equation in different material layers and sketch out a treatment to increase computational efficiency based on decomposition methods.

6.5.2.1   Simplified Models for the Crystal Growth Apparatus: Convection-Reaction Equation for the Simulation of the Crystal Growth

We apply the operator-splitting methods for the convection-reaction equation; see further numerical results in [88] and [89].

We deal with a first-order partial differential equation given as a transport equation in the following:

$$\partial_t u_1 = -v_1 \partial_x u_1 - \lambda u_1, \text{ for } (x,t) \in [0,X] \times [0,T], \quad (6.67)$$

$$\partial_t u_2 = -v_2 \partial_x u_2 + \lambda u_1, \text{ for } (x,t) \in [0,X] \times [0,T], \quad (6.68)$$

$$u_1(x,0) = \begin{cases} 1, & 0.1 \le x \le 0.3 \\ 0, & \text{otherwise} \end{cases}, \quad (6.69)$$

$$u_2(x,0) = 0, \text{ for } x \in [0,X], \quad (6.70)$$

$$u_1(0,t) = u_2(0,t) = 0, \text{ for } t \in [0,T], \quad (6.71)$$

where $\lambda \in \mathbb{R}^+$ and $v_1, v_2 \in \mathbb{R}^+$. We have the time interval $t \in [0,T]$ and the space interval $x \in [0,X]$. We apply the equations with $X = 1.5$ and $T = 1.0$. We rewrite the equation system (6.67)–(6.71) in operator notation, and end up with the following equations:

$$\partial_t u = Au + Bu, \quad (6.72)$$

$$u(x,0) = \begin{cases} (1,0)^T, & 0.1 \le x \le 0.3 \\ (0,0)^T, & \text{otherwise} \end{cases},$$

where $u = (u_1, u_2)^T$.
Our split operators are

$$A = \begin{pmatrix} -v_1 \partial_x & 0 \\ 0 & -v_2 \partial_x \end{pmatrix}, \quad B = \begin{pmatrix} -\lambda & 0 \\ \lambda & 0 \end{pmatrix}. \quad (6.73)$$

We use the finite difference method as a spatial discretization method and solve the time discretization analytically.

For the spatial discretization, we consider the interval $x \in [0, 1.5]$ uniformly partitioned with a step size of $\Delta x = 0.1$. For the transport term, we use an upwind finite difference discretization given as

$$\partial_x u_i = \frac{u_i - u_{i-1}}{\Delta x}. \quad (6.74)$$

We use the given impulses for the initial values:

$$u_1(x) = \begin{cases} 1, & 0.1 \le x \le 0.3 \\ 0, & \text{otherwise} \end{cases}, \quad (6.75)$$

and

$$u_2(x) = 0, \ x \in [0, 1.5]. \quad (6.76)$$

We deal with two indices for the discretized equation for the iterative operator-splitting method and its application to our transport equation. Index i denotes the spatial discretization, and index j denotes the number of iteration steps.

We first solve all the equations with index i, that means all 16 equations for each point. We then apply our iteration steps to arrive at the first time-step. We finish for one time partition, and we repeat this process four times more for the computations of five partitions, an so forth.

In the following equations, we write the iterative operator-splitting algorithm by taking into account the discretization in space. The discretization in time is solved analytically. For the time interval $[t^n, t^{n+1}]$, we solve the following problems consecutively for $j = 1, 3, 5, \ldots$. The split approximation at the time level $t = t^{n+1}$ is defined as $u_i^{n+1} \equiv u_{i,iter}(t^{n+1})$.

We have the following algorithm:

$$\partial_t u_{1,i,j} = -v_1/\Delta x(u_{1,i,j} - u_{1,i-1,j}) - \lambda u_{1,i,j-1}, \tag{6.77}$$

$$\partial_t u_{2,i,j} = -v_2/\Delta x(u_{2,i,j} - u_{2,i-1,j}) + \lambda u_{1,i,j-1}, \tag{6.78}$$

$$\partial_t u_{1,i,j+1} = -v_1/\Delta x(u_{1,i,j} - u_{1,i-1,j}) - \lambda u_{1,i,j+1}, \tag{6.79}$$

$$\partial_t u_{2,i,j+1} = -v_2/\Delta x(u_{2,i,j} - u_{2,i-1,j}) + \lambda u_{1,i,j+1}, \tag{6.80}$$

$$u_{1,i,j}(0) = \begin{cases} 1, & \text{for } i = 1, 2, 3 \\ 0, & \text{otherwise} \end{cases}, \tag{6.81}$$

$$u_{2,i,j}(0) = 0, \text{ for } i = 0, \ldots, 15, \tag{6.82}$$

where $\lambda = 0.5$ and $v_1 = 0.5$ and $v_2 = 1.0$. For the time interval we use $t \in [0, 1]$.

The analytical solution of the equation system (6.77)–(6.82) is

$$u_1(x, t) = \begin{cases} \exp(-\lambda t), & \text{for } 0.1 + v_1 t \leq x \leq 0.3 + v_1 t \\ 0, & \text{otherwise} \end{cases}, \tag{6.83}$$

and

$$u_2(x, t) = \lambda(L_{1,2} + L_{2,2} + M_{12,2}), \tag{6.84}$$

$$L_{1,2} = \begin{cases} -\frac{1}{\lambda} \exp(-\lambda t), & \text{for } 0.1 + v_1 t \leq x \leq 0.3 + v_1 t \\ 0, & \text{otherwise} \end{cases}, \tag{6.85}$$

$$L_{2,2} = \begin{cases} \frac{1}{\lambda}, & \text{for } 0.1 + v_2 t \leq x \leq 0.3 + v_2 t \\ 0, & \text{otherwise} \end{cases}, \tag{6.86}$$

$$M_{12,2} = \begin{cases} \frac{1}{\lambda} \exp(-\lambda t), & \text{for } 0.1 + v_1 t \leq x \leq 0.1 + v_2 t \\ -\frac{1}{\lambda} \exp(-\lambda t) \cdot \\ \exp\left(-\left(\frac{\lambda}{v_1 - v_2}\right) \cdot \\ (x - v_1 t - 0.3)\right), & \text{for } 0.3 + v_1 t \leq x \leq 0.3 + v_2 t \\ 0, & \text{otherwise} \end{cases}. \tag{6.87}$$

For the end time $t_{end} = 1$, we check the results for the end point $x_1 = v_1 t + 0.3$. We obtain the exact solution of our equation:

$$u_1(x_1, t_{end}) = 0.60653, \quad u_2(x_1, t_{end}) = 0.$$

In Table 6.39 we present the errors, computed with the L_∞-norm, for the numerical and exact solutions at the end time $t = 1$ and end point $x = v_1 t + 0.3 = 0.8$. The experiments are programmed in MATLAB 7.0 and computed with a 2 GHz Linux personal computer with a calculation accuracy of about 10^{-313}.

TABLE 6.39: Numerical results for the first example using iterative splitting method.

Number of time partitions n	Iteration steps j	err_{u_1, L_∞}	err_{u_2, L_∞}
1	2	2.679116×10^{-1}	2.465165×10^{-1}
1	4	1.699365×10^{-1}	3.584424×10^{-1}
1	10	2.702681×10^{-2}	5.327567×10^{-2}
5	2	2.472959×10^{-1}	6.812055×10^{-2}
5	4	1.181408×10^{-1}	4.757047×10^{-2}
5	10	1.680711×10^{-2}	1.496981×10^{-2}
10	2	2.289850×10^{-1}	4.246663×10^{-2}
10	4	1.121958×10^{-1}	2.498364×10^{-2}
10	10	8.999232×10^{-3}	2.819985×10^{-3}

6.5.2.2 Simplified Models for the Crystal Growth Apparatus: Heat Equation with Nonlinear Heat Source

In the second test example, we deal with a two-dimensional heat equation with a nonlinear heat source, see [3].

The two-dimensional heat equation is given as

$$\partial_t u(x, y, t) = u_{xx} + u_{yy}$$
$$-4(1 + y^2) \exp(-t) \exp(x + y^2), \text{ in } \Omega \times [0, T], \quad (6.88)$$
$$u(x, y, 0) = \exp(x + y^2), \text{ on } \Omega, \quad (6.89)$$
$$u(x, y, t) = \exp(-t) \exp(x + y^2), \text{ on } \partial\Omega \times [0, T], \quad (6.90)$$

where the domain is $\Omega = [-1, 1] \times [-1, 1]$ and the end time is $T = 1$.

The exact solution is given as

$$u(x, y, t) = \exp(-t) \exp(x + y^2). \quad (6.91)$$

We choose the time interval $[0,1]$ and again apply the finite differences for space with $\Delta x = \Delta y = 2/19$. We use the operator-splitting method for decomposition. The domain is split into two subdomains.

The operator equation is given as

$$\partial_t u(x, y, t) = Au + Bu + f(x, y),$$

where $f(x, y) = -4(1 + y^2) \exp(-t) \exp(x + y^2)$, and the operators are given as

$$Au = \begin{cases} u_{xx} + u_{yy}, & \text{for } (x, y) \in \Omega_1 \\ 0, & \text{for } (x, y) \in \Omega_2 \end{cases}, \tag{6.92}$$

$$Bu = \begin{cases} 0, & \text{for } (x, y) \in \Omega_1 \\ u_{xx} + u_{yy}, & \text{for } (x, y) \in \Omega_2 \end{cases}, \tag{6.93}$$

with $\Omega_1 \cup \Omega_2 = \Omega$ and $\Omega_1 \cap \Omega_2 = \emptyset$.

The initial and boundary conditions are given as in Equations (6.88)–(6.90).

We choose the splitting intervals $\Omega_1 = [-1, 0] \times [-1, 1]$ and $\Omega_2 = [0, 1] \times [-1, 1]$.

For the approximation error we apply the maximum norm (L_∞-norm), given as $\text{err}_{u, L_\infty} = \max_{i,j} |u_{exact}(x_i, y_j, T) - u_{approx}(i\Delta x, j\Delta y, T)|$, where $T = 1.0$.

The results of the experiment are presented in Table 6.40. The experiments are programmed in MATLAB 7.0 and computed with a 2 GHz Linux personal computer, with computation accuracy of around 10^{-313}.

The maximal error decreases with increasing iteration steps. Further comparisons to other methods are shown in [88], and based on these results the iterative operator-splitting method is more accurate.

Graphically, the solution smooths with more relaxation steps as seen in Figure 6.18.

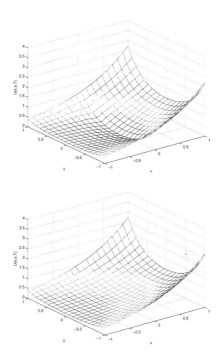

FIGURE 6.18: The numerical results of the second example after 10 iterations (left) and 20 iterations (right).

TABLE 6.40: Numerical results for the second example with iterative operator-splitting method and BDF3 method.

Iteration steps	Number of splitting partitions	err_{u,L_∞}
1	1	2.7183e+000
2	1	8.2836e+000
3	1	3.8714e+000
4	1	2.5147e+000
5	1	1.8295e+000
10	1	6.8750e-001
15	1	2.5764e-001
20	1	8.7259e-002
25	1	2.5816e-002
30	1	5.3147e-003
35	1	2.8774e-003

6.5.2.3 Real-Life Problem: Crystal Growth Apparatus

We concentrate on the stationary heat conduction in potentially anisotropic materials.

We consider the following underlying equation:

$$- \operatorname{div}(K_m(\theta) \nabla \theta) = f_m, \text{ in } \Omega_m, \quad (m \in M), \qquad (6.94)$$

where $\theta \geq 0$ represents the absolute temperature, the symmetric and positive definite matrix K_m represents the thermal conductivity tensor in the material m, $f_m \geq 0$ represents the heat sources in the material m due to some heating mechanism (e.g., induction or resistance heating), Ω_m is the domain of the material m, and M is a finite index set. We consider the case in which the thermal conductivity tensor is a diagonal matrix with temperature-independent anisotropy that is,

$$K_m(\theta) = \left(\kappa_{i,j}^m(\theta)\right), \quad \text{where} \quad \kappa_{i,j}^m(\theta) \begin{cases} \alpha_i^m \, \kappa_{\text{iso}}^m(\theta), & \text{for } i = j \\ 0, & \text{for } i \neq j \end{cases}, \qquad (6.95)$$

$\kappa_{\text{iso}}^m(\theta) > 0$ being the potential temperature-dependent thermal conductivity of the isotropic case, and $\alpha_i^m > 0$ being the anisotropy coefficients. As an example, the growth apparatus used in silicon carbide single crystal growth by physical vapor transport (PVT) is usually insulated by graphite felt in which the fibers are aligned in one particular direction, resulting in a thermal conductivity tensor of the form (6.95). We apply the finite volume scheme, based on control volumes, see [96], and consider the anisotropy in the thermal insulation of the PVT growth apparatus.

The temperature θ is assumed to be continuous throughout the entire domain $\overline{\Omega}$. The continuity of the normal component of the heat flux on the interface between different materials m_1 and m_2, $m_1 \neq m_2$, yields the following interface conditions, coupled with the heat equations (6.94):

$$\left(K_{m_1}(\theta) \, \nabla \theta\right)\restriction_{\overline{\Omega}_{m_1}} \cdot n_{m_1} = \left(K_{m_2}(\theta) \, \nabla \theta\right)\restriction_{\overline{\Omega}_{m_2}} \cdot n_{m_1}, \text{ on } \overline{\Omega}_{m_1} \cap \overline{\Omega}_{m_2}, \quad (6.96)$$

where $\restriction$ denotes restriction, and n_{m_1} denotes the unit normal vector pointing from material m_1 to material m_2.

We consider two types of outer boundary conditions, namely Dirichlet and Robin conditions. To that end, we decompose $\partial\Omega$ according to

- Let Γ_{Dir} and Γ_{Rob} be relatively open polyhedral subsets of $\partial\Omega$, such that $\partial\Omega = \overline{\Gamma}_{\text{Dir}} \cup \overline{\Gamma}_{\text{Rob}}$, $\Gamma_{\text{Dir}} \cap \Gamma_{\text{Rob}} = \emptyset$.

The boundary conditions are then given as

$$\theta = \theta_{\text{Dir}}, \text{ on } \overline{\Gamma}_{\text{Dir}}, \quad (6.97a)$$

$$-\left(K_m(\theta) \, \nabla \theta\right) \cdot n_m = \xi_m \left(\theta - \theta_{\text{ext},m}\right),$$
$$\text{on } \Gamma_{\text{Rob}} \cap \partial\Omega_m, \, m \in M, \quad (6.97b)$$

where n_m is the outer unit normal to Ω_m, $\theta_{\text{Dir}} \geq 0$ is the given temperature on Γ_{Dir}, $\theta_{\text{ext},m} \geq 0$ is the given external temperature ambient to $\Gamma_{\text{Rob}} \cap \partial\Omega_m$, and $\xi_m > 0$ is a transition coefficient.

Our apparatus is given in Figure 6.19.

The radius is 12 cm and the height is 45.3 cm. This domain represents a growth apparatus used in silicon carbide single crystal growth by the PVT method. The domain Ω consists of six subdomains Ω_m, $m \in \{1, \ldots, 6\}$, representing the insulation materials, graphite crucible, SiC crystal seed, gas enclosure, SiC powder source, and quartz. In order to use realistic functions for the isotropic parts $\kappa_{\text{iso}}^m(\theta)$ of the thermal conductivity tensors (cf. (6.95)), for gas enclosure, graphite crucible, insulation, and SiC crystal seed, we use the functions given by (A.1), (A.3b), (A.4b), and (A.7b) in [139]; for $\kappa_{\text{iso}}^5(\theta)$ (SiC powder source), we use [137, (A.1)], and for $\kappa_{\text{iso}}^6(\theta)$ (quartz), we use

$$\kappa_{\text{iso}}^6(\theta) = \left(1.82 - 1.21 \cdot 10^{-3} \frac{\theta}{\text{K}} + 1.75 \cdot 10^{-6} \frac{\theta^2}{\text{K}^2}\right) \frac{\text{W}}{\text{m K}}. \quad (6.98)$$

Hence, all functions $\kappa_{\text{iso}}^m(\theta)$ depend nonlinearly on θ. As mentioned in the introduction, the thermal conductivity in the insulation is typically anisotropic in PVT growth apparatus. In the numerical experiments reported below, we therefore vary the anisotropy coefficients (α_r^1, α_z^1) of the insulation while keeping $(\alpha_r^m, \alpha_z^m) = (1,1)$ for all other materials $m \in \{2, \ldots, 5\}$.

Heat sources $f_m \neq 0$ are supposed to be present *only* in the part of Ω_2 (graphite crucible) labeled "uniform heat sources" in the left-hand picture in Figure 6.20, satisfying $5.4 \text{ cm} \leq r \leq 6.6 \text{ cm}$ and $9.3 \text{ cm} \leq z \leq 42.0 \text{ cm}$.

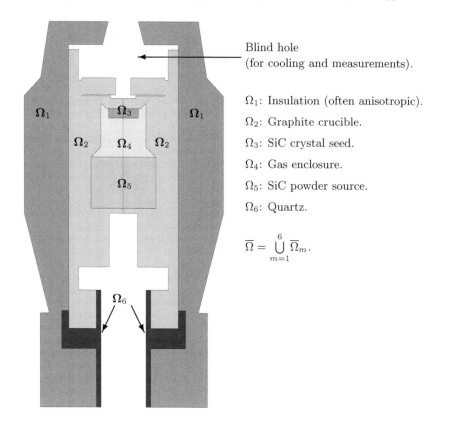

Blind hole
(for cooling and measurements).

Ω_1: Insulation (often anisotropic).

Ω_2: Graphite crucible.

Ω_3: SiC crystal seed.

Ω_4: Gas enclosure.

Ω_5: SiC powder source.

Ω_6: Quartz.

$$\overline{\Omega} = \bigcup_{m=1}^{6} \overline{\Omega}_m.$$

FIGURE 6.19: The underlying apparatus given as an axis-symmetric domain with the different material regions.

In that region, f_2 is set to be the constant value $f_2 = 1.23$ MW/m^3, which corresponds to a total heating power of 1.8 KW. This serves as an approximation to the situation typically found in a radio-frequent induction-heated apparatus, where a moderate skin effect concentrates the heat sources within a few millimeters of the conductor's outer surface.

Here, our main goal is to illustrate the effectiveness of our spatial discretization method based on finite volume schemes. With such results, we can apply our decomposition methods to separate the anisotropic directions.

This is the case if the anisotropy in the thermal conductivity of the insulation is sufficiently large. For such a case we expect the isotherms to be almost parallel to the direction with the larger anisotropy coefficient. Because using the Dirichlet boundary condition (6.97a) can suppress such an alignment of the isotherms, we prefer the Robin condition (6.97b) on all of $\partial\Omega$ instead. For $m \in \{1, 2, 6\}$, we set $\theta_{\text{ext},m} = 500$ K and $\xi_m = 80$ W/(m^{2}K).

We now present the results of our numerical experiments, varying the anisotropy coefficients (α_r^1, α_z^1) in the insulation.

In each case, we use a fine grid consisting of 61,222 triangles. We start with the isotropic case $(\alpha_r^1, \alpha_z^1) = (1,1)$ depicted on the right-hand side of Figure 6.20. Figure 6.21 shows the computed temperature fields for the moderately anisotropic cases $(\alpha_r^1, \alpha_z^1) = (10,1)$ (left), $(\alpha_r^1, \alpha_z^1) = (1,10)$ (middle), $(\alpha_r^1, \alpha_z^1) = (10,1)$ in top and bottom insulation parts, $(\alpha_r^1, \alpha_z^1) = (1,10)$ in insulation side wall (right).

The maximal temperatures established in the four experiments are collected in Table 6.41.

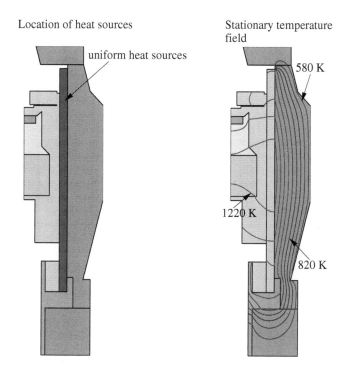

Location of heat sources

uniform heat sources

Stationary temperature field

580 K

1220 K

820 K

FIGURE 6.20: Left: Location of the heat sources; Right: Computed temperature field for the isotropic case $\alpha_r^1 = \alpha_z^1 = 1$, where the isotherms are spaced at 80 K.

Comparing the temperature fields in Figures 6.20 and 6.21 and the maximal temperatures listed in Table 6.41, we find that any anisotropy reduces the effectiveness of the thermal insulation, where a stronger anisotropy results in less insulation. A stronger anisotropy results in a less effective insulation and the value above 1 improves the insulation's thermal conductivity in that

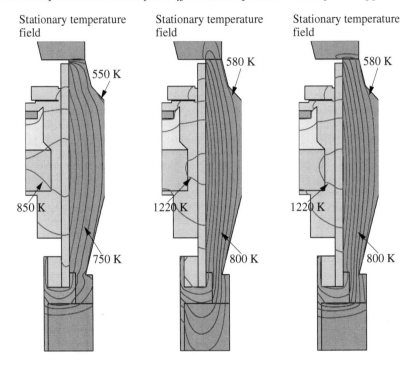

FIGURE 6.21: Computed temperature fields for the moderately anisotropic cases $(\alpha_r^1, \alpha_z^1) = (10, 1)$ (left, isotherms spaced at 50 K); $(\alpha_r^1, \alpha_z^1) = (1, 10)$ (middle, isotherms spaced at 80 K); $(\alpha_r^1, \alpha_z^1) = (10, 1)$ in top and bottom insulation parts, $(\alpha_r^1, \alpha_z^1) = (1, 10)$ in insulation side wall (right, isotherms spaced at 80 K).

direction. Similarly, when reducing one of the anisotropy coefficients to a value below 1, a stronger anisotropy would result in improved insulation.

The application of decoupling methods to this real-life application is very important because of its complicated domains with different parameters. Each simpler domain can be computed more accurately and parallel computation is possible. In this version, only the test examples are performed with the decomposition methods.

REMARK 6.16 The underlying real-life application can be computed more efficiently with spatial decomposition methods. In further experiments, we will show that spatial decomposition methods are more adaptive to the underlying domains and can benefit from parallelization. ▯

TABLE 6.41: Maximal temperatures for numerical experiments, depending on the anisotropy coefficients (α_r^1, α_z^1) of the insulation (cf. Figures 6.20 - 6.21).

α_r^1	α_z^1	Maximal temperature [K]
1	1	1273.18
1	10	1232.15
1-10, mixed	1-10, mixed	1238.38
10	1	918.35

6.5.3 Elastic Wave Propagation

In these sections, we focus on decoupling the wave equations motivated by a realistic problem involving seismic sources and waves. The complicated wave propagation process is decomposed into simpler processes with respect to the directions of the propagation. We discretize and solve the simpler equations by more accurate higher-order methods, see [103].

With the test examples, we verify the decomposition methods for the linear acoustic wave equations. Moreover, the delicate initialization of the right-hand side, while using a Dirac function, is presented with respect to the decomposition method. The benefit of spatial splitting methods, applied to the multidimensional operators, is discussed in several examples.

6.5.3.1 Real-Life Application of the Elastic Wave Propagation

In this application, we refer to our underlying model equations presented in Section 1.3.3. Based on these equations we discuss the discretization and decomposition.

We focus on a higher-order splitting method for a real-life application problem involving the approximations of time and space; we also focus on how the Dirac function is approximated.

During numerical testing we have observed a need to reduce the allowable time-step in the case where the ratio of λ over μ becomes too large. This is likely from the influence of the explicitly treated mixed derivative. For really high ratios (> 20) a reduction of 35% was necessary to avoid numerical instabilities.

6.5.3.2 Initial Values and Boundary Conditions

In order to start the time-stepping scheme we need to know the values at two earlier time levels. Starting at time $t = 0$, we know the value at level $n = 0$ as $U^0 = g_0$. The value at level $n = -1$ can be obtained by Taylor expansion as shown below:

$$U^{-1} = U^0 - \tau \partial_t U^0 + \frac{\tau_n^2}{2} \partial_{tt} U^0 - \frac{\tau^3}{6} \partial_{ttt} U^0 + \frac{\tau^4}{24} \partial_{tttt} U^0 + \mathcal{O}(\tau^5), \quad (6.99)$$

where we use

$$\partial_t U_{j,k}^0 = g_{1j,k}, \quad (6.100)$$

$$\partial_{tt} U_{j,k}^0 \approx \frac{1}{\rho} \left(\mathcal{M}_4 g_{0j,k} \right) + f_{j,k} \right), \quad (6.101)$$

$$\partial_{ttt} U_{j,k}^0 \approx \frac{1}{\rho} \left(\mathcal{M}_4 g_{1j,k} \right) + \partial_t f^0{}_{j,k} \right), \quad (6.102)$$

$$\partial_{tttt} U^0{}_{j,k} \approx \frac{1}{\rho} \left(\mathcal{M}_2^2 g_{0j,k} \right) + \mathcal{M}_4 f^0{}_{j,k} + \partial_{tt} f^0{}_{j,k} \right), \quad (6.103)$$

and also for (6.102) and (6.103).

The approximation of our right-hand side is given as

$$\partial_t f^0{}_{j,k} \approx \frac{f^1{}_{j,k} - f^{-1}{}_{j,k}}{2\tau}, \quad (6.104)$$

$$\partial_{tt} f^0{}_{j,k} \approx \frac{f^1{}_{j,k} - 2 f^0{}_{j,k} + f^{-1}{}_{j,k}}{\tau_n^2}. \quad (6.105)$$

We are not considering the boundary value problem and thus we will not be concerned with constructing proper difference stencils at grid points near the boundaries of the computational domain. We have simply added a two-point-thick layer of extra grid points at the boundaries of the domain and assigned the correct analytical solution at all points in the layer for every time-step.

REMARK 6.17 For the Dirichlet boundary conditions, the splitting method, see (4.35)–(4.37), also conserves the conditions. For the three equations (i.e., for U^*, U^{**} and for U^{n+1}), we can use the same conditions.

For the Neumann boundary conditions and other boundary conditions of higher order, we also have to split the boundary conditions with respect to the split operators, see [154]. □

6.5.3.3 Test Example of the Two-Dimensional Wave Equation

The first test example is a two-dimensional example. The splitting method is presented in Section 4.2.3 as well as the model equations.

We apply in the first test case a forcing function given as

$$\begin{aligned} f = \big(&\sin(t-x)\sin(y) - 2\mu \sin(t-x)\sin(y) \\ &-(\lambda+\mu)(\cos(x)\cos(t-y) + \sin(t-x)\sin(y)), \\ &\sin(t-y)\sin(x) - 2Vs^2 \sin(x)\sin(t-y) \\ &-(\lambda+\mu)(\cos(t-x)\cos(y) + \sin(y)\sin(t-y)) \big)^T, \quad (6.106) \end{aligned}$$

giving the analytical solution

$$U^{\text{true}} = \big(\sin(x - t)\sin(y),\ \sin(y - t)\sin(x) \big)^T. \qquad (6.107)$$

Using the splitting method, we solved (1.10) on a domain $\Omega = [-1, 1] \times [-1, 1]$ and time interval $t \in [0, 2]$. We used two sets of material parameters; for the first case ρ, λ, and μ were all equal to 1, for the second case ρ and μ were 1 and λ was set to 14. Solving on four different grids with a refinement factor of two in each direction between the successive grids, we obtained the results shown in Table 6.42. For all test examples, the equidistant time-step is given as $\tau = 0.0063$. The errors are measured in the L_∞-norm defined as $\|U_{j,k}\| = \max\left(\max_{j,k} |u_{j,k}|, \max_{j,k} |v_{j,k}|\right)$.

TABLE 6.42: Errors in max-norm for decreasing h and smooth analytical solution U^{true}. Convergence rate indicates fourth-order convergence for the split scheme.

Grid step	Time $t = 2$, $e_h = \text{err}_{U,L_\infty} = \|U^n - U^{\text{true}}\|_\infty$			
h	case 1	$\log_2(\frac{e_{2h}}{e_h})$	case 2	$\log_2(\frac{e_{2h}}{e_h})$
0.05	1.7683e-07		2.5403e-07	
0.025	1.2220e-08	3.855	2.1104e-08	3.589
0.0125	7.9018e-10	3.951	1.4376e-09	3.876
0.00625	5.0013e-11	3.982	9.2727e-11	3.955

As can be seen, we get the expected fourth-order convergence for problems with smooth solutions.

To check the influence of the splitting error $\mathcal{N}_{4,\theta}$ on the error, we solved the same problems using the nonsplit scheme (4.34). The results are shown in Table 6.43. The errors are only marginally smaller than for the split scheme.

6.5.3.4 Singular Forcing Terms

In seismology and acoustics, it is common to use spatial singular forcing terms that can look like

$$f = F\delta(x)g(t), \qquad (6.108)$$

where F is a constant direction vector. A numeric method for Equation (1.10), see Chapter 1, needs to approximate the Dirac function correctly in order to achieve full convergence. Obviously, we cannot expect convergence close to the source as the solution will be singular for two- and three-dimensional domains.

The analyses in [188] and [195] demonstrate that it is possible to derive regularized approximations of the Dirac function that result in a point-wise

TABLE 6.43: Errors in max-norm for decreasing h and smooth analytical solution U^{true} and using the non-split scheme. Comparing with Table 6.42 we see that the splitting error is very small for this case.

Grid step h	Time $t = 2$, $e_h = \mathrm{err}_{U,L_\infty} = \|U^n - U^{\mathrm{true}}\|_\infty$	
	case 1	case 2
0.05	1.6878e-07	2.4593e-07
0.025	1.1561e-08	2.0682e-08
0.0125	7.4757e-10	1.4205e-09
0.00625	4.8112e-11	9.2573e-11

convergence of the solution away from the sources. Based on these analyses, we define one second-order (δ_{h^2}) and one fourth-order (δ_{h^4}) regularized approximation of the one-dimensional Dirac function:

$$\delta_{h^2}(\tilde{x}) = \frac{1}{h} \begin{cases} 1 + \tilde{x}, & -h \le \tilde{x} < 0, \\ 1 - \tilde{x}, & 0 \le \tilde{x} < h, \\ 0, & \text{elsewhere,} \end{cases} \tag{6.109}$$

$$\delta_{h^4}(\tilde{x}) = \frac{1}{h} \begin{cases} 1 + \frac{11}{6}\tilde{x} + \frac{5}{8}\tilde{x}^2 + \frac{1}{6}\tilde{x}^3, & -2h \le \tilde{x} < -h, \\ 1 + \frac{1}{2}\tilde{x} - \tilde{x}^2 - \frac{1}{2}\tilde{x}^3, & -h \le \tilde{x} < 0, \\ 1 - \frac{1}{2}\tilde{x} - \tilde{x}^2 + \frac{1}{2}\tilde{x}^3, & 0 \le \tilde{x} < h, \\ 1 - \frac{11}{6}\tilde{x} + \tilde{x}^2 - \frac{1}{6}\tilde{x}^3, & h \le \tilde{x} < 2h, \\ 0, & \text{elsewhere,} \end{cases} \tag{6.110}$$

where $\tilde{x} = x/h$. The two- and three-dimensional Dirac functions are then approximated as $\delta_{h^{2,4}}(\tilde{x})\delta_{h^{2,4}}(\tilde{y})$ and $\delta_{h^{2,4}}(\tilde{x})\delta_{h^{2,4}}(\tilde{y})\delta_{h^{2,4}}(\tilde{z})$. The chosen time dependence was a smooth function given by

$$g(t) = \begin{cases} \exp(-1/(t(1-t))), & 0 \le t < 1, \\ 0, & \text{elsewhere,} \end{cases} \tag{6.111}$$

which is C^∞. Using this forcing function, we can compute the analytical solution by integrating the Green's function given in [61]. The integration was done using numerical quadrature routines from MATLAB. Figures 6.22 and 6.23 show examples of what the errors look like on a radius passing through the singular source at time $t = 0.8$ for different grid sizes h and the two approximations δ_{h^2} and δ_{h^4}. We see that the error is smooth and converges a small distance away from the source. However, using δ_{h^2} limits the convergence to second-order, while using δ_{h^4} gives the full fourth-order convergence away from the singular source. When $t > 1$, the forcing goes to zero and the solution will be smooth everywhere. Table 6.44 shows the convergence behavior at time $t = 1.1$ for four different grids. Note that the full

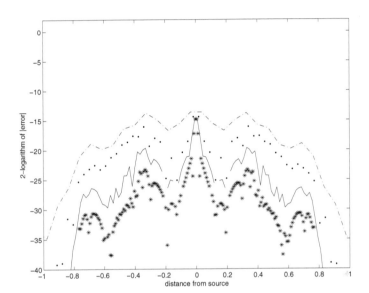

FIGURE 6.22:   The 2-logarithm of the error along a line going through
the source point for a point force located at $x = 0, y = 0$, and approximated
in space by (6.110). Note that the error decays with $\mathcal{O}(h^4)$ away from the
source but not near it. The grid sizes were
$h = 0.05\,(-\cdot)$, $0.025\,(\cdot)$, $0.0125\,(-)$, $0.00625\,(*)$. The numerical quadrature
had an absolute error of approximately $10^{-11} \approx 2^{-36}$, so the error cannot be
resolved beneath that limit.

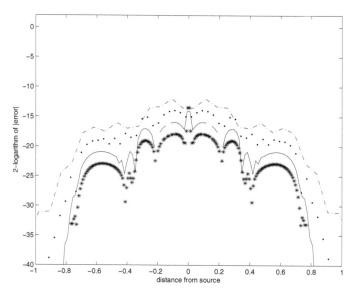

FIGURE 6.23: The 2-logarithm of the error along a line going through the source point for a point force located at $x = 0, y = 0$, and approximated in space by (6.109). Note that the error only decays with $\mathcal{O}(h^2)$ away from the source. The grid sizes were $h = 0.05\,(-\cdot),\ 0.025\,(\cdot),\ 0.0125\,(-),\ 0.00625\,(*)$.

TABLE 6.44: Errors in max-norm for decreasing h and analytical solution U^{true}. Convergence rate approaches fourth-order after the singular forcing term goes to zero of the two-dimensional split scheme.

Grid step h	Time $t = 1.1$, $e_h = \text{err}_{U,L_\infty} = \lVert U^n - U^{\text{true}} \rVert_\infty$ case 1	$\log_2\left(\frac{e_{2h}}{e_h}\right)$
0.05	1.1788e-04	
0.025	1.4146e-05	3.0588
0.0125	1.3554e-06	3.3836
0.00625	1.0718e-07	3.6606
0.003125	7.1890e-09	3.8981

convergence is achieved even if the lower-order δ_{h^2} is used as an approximation for the Dirac function.

The convergence rate approaches four as we refine the grids, even though the solution was singular up to time $t = 1$.

We implement the two-dimensional case in MATLAB on a single CPU. Further computations are also done on multiple CPUs with optimal speed.

The spatial grid steps are given as $h = \Delta x = \Delta y = \Delta y \in \{0.05, 0.025, 0.0125, 0.00625, 0.003125\}$, and the time-step is given as $\tau = 0.0063$. The material parameters for the elastic wave propagation equations are $\lambda = 1$, $\mu = 1$, and $\rho = 1$. The total number of grid points are about 26,000.

A visualization of the elastic wave propagation is given in Figures 6.24 and 6.25.

To visualize the influence of the singular point force for the two-dimensional computation we use the x-component of the solution at time $t = 1$ over the normed domain $[-1,1] \times [-1,1]$.

6.5.3.5 Computational Cost of the Splitting Method

For a two-dimensional problem, the fourth-order explicit method (4.33) can be implemented using approximately 160 *floating point operations* (flops) per grid point.

The splitting method requires approximately 120 flops (first step) plus two times 68 flops (second and third step) for a total of 256 flops. This increase of about 60% in the number of flops is somewhat offset by the larger time-steps allowed by the splitting method, especially for "nice" material properties, making the two methods roughly comparable in computational cost.

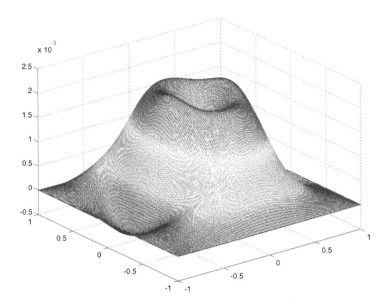

FIGURE 6.24:   The x-component of the solution for a singular point force at time $t = 1$ and spatial grid step $h = 0.0125$.

6.5.3.6   A Three-Dimensional Splitting Method

As the splitting method discussed for two dimensions in Section 4.2.3, we extend the splitting method to three dimensions in the applications. Therefore, a first discussion of the discretization and decomposition of the operator is important. In three dimensions, a fourth-order difference approximation of the operator becomes

$$
\mathcal{M}_4 = \left(
\begin{array}{c}
(\lambda + 2\mu)\left(1 - \frac{h^2}{12}D^{x2}\right)D^{x2} + \mu\left(1 - \frac{h^2}{12}D^{y2}D^{y2} + 1 - \frac{h^2}{12}D^{z2}\right)D^{z2} \\
(\lambda + \mu)\left(1 - \frac{h^2}{6}D^{x2}\right)D_0^x\left(1 - \frac{h^2}{6}D^{y2}\right)D_0^y \\
(\lambda + \mu)\left(1 - \frac{h^2}{6}D^{x2}\right)D_0^x\left(1 - \frac{h^2}{6}D^{z2}\right)D_0^z
\end{array}
\right.
$$

$$
(\lambda + \mu)\left(1 - \frac{h^2}{6}D^{x2}\right)D_0^x\left(1 - \frac{h^2}{6}D^{y2}\right)D_0^y
$$
$$
(\lambda + 2\mu)\left(1 - \frac{h^2}{12}D^{y2}\right)D^{y2} + \mu\left(1 - \frac{h^2}{12}D^{x2}\right)D^{x2} + \mu\left(1 - \frac{h^2}{12}D^{z2}\right)D^{z2}
$$
$$
(\lambda + \mu)\left(1 - \frac{h^2}{6}D^{z2}\right)D_0^z\left(1 - \frac{h^2}{6}D^{y2}\right)D_0^y
$$

$$
\left.
\begin{array}{c}
(\lambda + \mu)\left(1 - \frac{h^2}{6}D^{x2}\right)D_0^x\left(1 - \frac{h^2}{6}D^{z2}\right)D_0^z \\
(\lambda + \mu)\left(1 - \frac{h^2}{6}D^{y2}\right)D_0^y\left(1 - \frac{h^2}{6}D^{z2}\right)D_0^z \\
(\lambda + 2\mu)\left(1 - \frac{h^2}{12}D^{z2}D^{z2} + \mu\left(1 - \frac{h^2}{12}D^{x2}\right)D^{x2} + 1 - \frac{h^2}{12}D^{y2}\right)D^{y2}
\end{array}
\right),
$$

operating on grid functions $U_{j,k,l}^n$ defined at grid points x_j, y_k, z_l, t_n similarly to the two-dimensional case. We can split $\mathcal{M}_4$ into six parts; $\mathcal{M}_{xx}, \mathcal{M}_{yy}, \mathcal{M}_{zz}$ containing the three second-order directional difference operators, and $\mathcal{M}_{xy}, \mathcal{M}_{yz}, \mathcal{M}_{xz}$ containing the mixed difference operators.

We could split this scheme in a number of different ways depending on how we treat the mixed derivative terms. We have chosen to implement the following split scheme in three dimensions:

1. $\rho \dfrac{U_{j,k,l}^* - 2U_{j,k,l}^n + U_{j,k,l}^{n-1}}{\tau_n^2} = \mathcal{M}_4 U_{j,k,l}^n + \theta f_{j,k,l}^{n+1} + (1 - 2\theta) f_{j,k,l}^n + \theta f_{j,k,l}^{n-1}),$

2. $\rho \dfrac{U_{j,k,l}^{**} - U_{j,k,l}^*}{\tau_n^2} = \theta \mathcal{M}_{xx} \left(U_{j,k,l}^{**} - 2U_{j,k,l}^n + U_{j,k,l}^{n-1} \right)$

$\qquad\qquad\qquad + \dfrac{\theta}{2} (\mathcal{M}_{xy} + \mathcal{M}_{xz}) \left(U_{j,k,l}^* - 2U_{j,k,l}^n + U_{j,k,l}^{n-1} \right),$

3. $\rho \dfrac{U_{j,k,l}^{***} - U_{j,k,l}^{**}}{\tau_n^2} = \theta \mathcal{M}_{xx} \left(U_{j,k,l}^{***} - 2U_{j,k,l}^n + U_{j,k,l}^{n-1} \right)$

$\qquad\qquad\qquad + \dfrac{\theta}{2} (\mathcal{M}_{xy} + \mathcal{M}_{yz}) \left(U_{j,k,l}^{**} - 2U_{j,k,l}^n + U_{j,k,l}^{n-1} \right),$

4. $\rho \dfrac{U_{j,k,l}^{n+1} - U_{j,k,l}^{***}}{\tau_n^2} = \theta \mathcal{M}_{xx} \left(U_{j,k,l}^{n+1} - 2U_{j,k,l}^n + U_{j,k,l}^{n-1} \right)$

$\qquad\qquad\qquad + \dfrac{\theta}{2} (\mathcal{M}_{xz} + \mathcal{M}_{yz}) \left(U_{j,k,l}^{***} - 2U_{j,k,l}^n + U_{j,k,l}^{n-1} \right).$

The properties such as splitting error, accuracy, stability, and so forth, for the three-dimensional case are similar to the two-dimensional case treated in the earlier sections.

6.5.3.7 Test Example of the Three-Dimensional Wave Equation

We have performed some numerical experiments with the three-dimensional scheme in order to test the convergence and stability. We apply our splitting scheme, presented in Section 6.5.3.6, and use a forcing

$$f = \big(-(-1 + \lambda + 4\mu) \sin(t - x) \sin(y) \sin(z) -$$
$$(\lambda + \mu) \cos(x)(2 \sin(t) \sin(y) \sin(z) + \cos(t) \sin(y + z)),$$
$$-(-1 + \lambda + 4\mu) \sin(x) \sin(t - y) \sin(z) -$$
$$(\lambda + \mu) \cos(y)(2 \sin(t) \sin(x) \sin(z) + \cos(t) \sin(x + z)),$$
$$-(\lambda + \mu) \cos(t - y) \cos(z) \sin(x) - \sin(y)((\lambda + \mu) \cos(t - x) \cos(z) +$$
$$(-1 + \lambda + 4\mu) \sin(x) \sin(t - z)) \big)^T, \tag{6.112}$$

giving the analytical solution

$$U^{\text{true}} = \big(\sin(x - t) \sin(y) \sin(z),$$
$$\sin(y - t) \sin(x) \sin(z),$$
$$\sin(z - t) \sin(x) \sin(y) \big)^T. \tag{6.113}$$

As in the earlier examples, we tested this for a number of different grid sizes. Using the same two sets of material parameters as for the two-dimensional case, we computed up until $t = 2$ and checked the maximum error for all components of the solution. The results are given in Table 6.45.

TABLE 6.45: Errors in max-norm for decreasing h and smooth analytical solution U^{true}. Convergence rate indicates fourth-order convergence of the three-dimensional split scheme.

h	$t = 2$, $e_h = \text{err}_{U,L_\infty} = \|U^n - U^{\text{true}}\|_\infty$			
	case 1	$\log_2\left(\frac{e_{2h}}{e_h}\right)$	case 2	$\log_2\left(\frac{e_{2h}}{e_h}\right)$
0.1	4.2986e-07		1.8542e-06	
0.05	3.5215e-08	3.61	1.3605e-07	3.77
0.025	3.0489e-09	3.53	8.0969e-09	4.07
0.0125	2.0428e-10	3.90	4.7053e-10	4.10

We implement the three-dimensional case in MATLAB on a single CPU. Further computations are also done on multiple CPUs with optimal speed.

The spatial grid steps are given as $h = \Delta x = \Delta y = \Delta y \in \{0.1, 0.05, 0.025, 0.0125\}$, and the time-step is given as $\tau = \tau = 0.0046$. The parameters for the elastic wave propagation are given as $\lambda = 14$, $\mu = 1$, and $\rho = 1$. We obtain a total number of grid points of about $1 \, 10^6$.

The initial solution is given as $U = 0$.

In Figure 6.25, we visualize the influence of the singular point force for the three-dimensional computation using the y-component on a plane. The solution is given at the time $t = 1$ and at the normed domain $[-1, 1] \times [-1, 1] \times [-1, 1]$. Further, the isosurfaces for the solution U are also visualized to show the wave fronts.

REMARK 6.18 Our splitting scheme has been shown to work well in practice for different types of material properties. It is comparable to the fully explicit fourth-order scheme (4.33) in terms of computational cost but should be easier to implement, as no difference approximations of higher-order operators are needed.

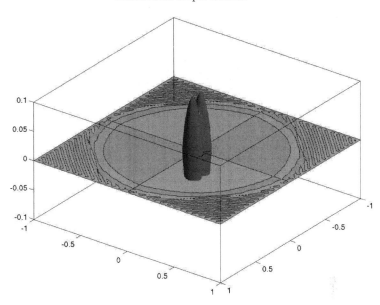

FIGURE 6.25: Contour plot of the y-component on a plane for the 3D case with a singular force. The inclusion of an isosurface for $||U||$ is also done.

A vital component of a model in seismology is the stable higher-order approximation of the boundary conditions, something we have saved for a future manuscript. ⬚

6.5.4 Magnetic Layers

The motivation for the study is coming from modeling ferromagnetic materials, used in data storage devices and laptop displays. The models are based on the Landau-Lifschitz-Gilbert equation, see [144], and are extended by the interaction of different magnetic layers. The application of such model problems include switch processes in magnetic transistors (FET) or data devices. Mathematically the equations cannot be solved analytically and numerical methods are important. Therefore, numerical methods for stable discretizations are important because of the known blow-up effects, see [19] and [174]. We apply the scalar theory and present the extention to more complicated systems of magnetic models. In the numerical examples we present first results of single layers and influence with external magnetic fields.

6.5.4.1 Discretization and Solver Methods

For the discretization method we apply a stable implicit method. In general the equations are iteratively solved with semi-discretization. We obtain a general discretized equation given as

$$\partial_t \, m_1^i = A_1 m_1^i + B_1 m_1^i + H_{eff,1}(m_1^i, m_2^{i-1}) \,, \tag{6.114}$$
$$\partial_t \, m_2^{i+1} = A_2 m_2^{i+1} + B_2 m_2^{i+1} + H_{eff,2}(m_1^i, m_2^{i+1}) \,, \tag{6.115}$$

where $i = 0, 1, 2, \dots$.

In the following we discuss the scalar and the system cases.

6.5.4.2 Implicit Discretization for Scalar LLG Equations

We apply the following discretization :
$V_h \subset W^{1,2}(\Omega, \mathbb{R}^3)$, $\mathcal{T}_h$ a triangulation, $m_h^0 \in V_h$, $m_h^j \in V_h$

The full discretization is given with a linearization :

$$(\partial_t m_h^{j+1}, \phi_h)_h + \alpha(m_h^j \times \partial_t \, m_h^{j+1}, \phi_h)_h \tag{6.116}$$
$$= (1 + \alpha^2)(\overline{m}_h^{j+1/2} \times \Delta_h \overline{m}_h^{j+1/2}, \phi_h)_h$$

where $\phi_h \in V_h$ and j : index for the time-steps.
$\partial m_h^{j+1} = k^{-1}(m_h^{j+1} - m_h^j)$ and $k = t^{j+1} - t^j$
$\overline{m}_h^{j+1/2} = 1/2(m_h^{j+1} + m_h^j)$

Further the constraint : $|m_h^j| = 1$ discrete energy law.

We deal with the anisotropic model and add the magnetic field H_{eff}.
In the weak formulation, we can formulate

$$(\partial_t \, m, \phi) + \alpha(m \times \partial_t m, \phi) = (1 + \alpha^2)(m \times \Delta m, \phi) \tag{6.117}$$
$$+ (1 + \alpha^2)(m \times H_{eff}, \phi) \,,$$

with the initial condition $m(0) = m_0 \in W^{1,2}(\Omega; S^2)$ and the magnetic field is given as H_{eff}.

The full discretization is given with a linearization:

$$(\partial_t m_h^{j+1}, \phi_h)_h + \alpha(m_h^j \times \partial_t \, m_h^{j+1}, \phi_h)_h \tag{6.118}$$
$$= (1 + \alpha^2)(\overline{m}_h^{j+1/2} \times (\Delta_h \overline{m}_h^{j+1/2} + \overline{H_{eff}}_h^{j+1/2}), \phi_h)_h$$

where $\phi_h \in V_h$ and j : index for the time-steps.

6.5.4.3 Stability Theory for the LLG Equation

In the previous work of [19] the problems of higher-order derivatives are discussed.

We will concentrate on the nonlinear case $H(u)$ and time-dependent case $H(t)$.

For both we could derive a stabile implicit discretization method. In the following we present the fixpoint-algorithm to solve Equation (6.118).

ALGORITHM 6.1
Set $m_h^{j+1,l} := m_h^j$ and $l := 0$.

(i) Compute $m_h^{j+1,l+1} \in V_h$ such that for all $\phi_h \in V_h$ there holds

$$1/k(m_h^{j+1,l+1}, \phi_h)_h + \alpha/k(m_h^j \times m_h^{j+1,l+1}, \phi_h)_h \qquad (6.119)$$
$$-(1+\alpha^2)/4(\overline{m}_h^{j+1,l+1} \times (\Delta_h \overline{m}_h^{j+1,l} + H_{eff}^{j+1}), \phi_h)_h$$
$$-(1+\alpha^2)/4(\overline{m}_h^j \times (\Delta_h \overline{m}_h^{j+1,l} + H_{eff}^{j+1}), \phi_h)_h$$
$$-(1+\alpha^2)/4(\overline{m}_h^j \times (\Delta_h \overline{m}_h^{j+1,l+1} + H_{eff}^{j+1}), \phi_h)_h$$
$$= 1/k(m_h^j, \phi_h)_h + (1+\alpha^2)/4(\overline{m}_h^j \times (\Delta_h \overline{m}_h^j + H_{eff}^j), \phi_h)_h$$

(ii) If $||m_h^{j+1,l+1} - m_h^{j+1,l}||_h \leq \epsilon$ then stop and set $m_h^{j+1} := m_h^{j+1,l+1}$.
(iii) Set $l := l+1$ and go to (i).

We can derive the error estimates.

LEMMA 6.1

Suppose that we have γ dependent from h, k, and α, and $|m_h^j(x_m)| = 1$ for all m. Then for all $l \geq 1$ there holds

$$||m_h^{j+1,l+1} - m_h^{l+1,l}||_h \leq \Theta\gamma||m_h^{j+1,l} - m_h^{l+1,l-1}||_h. \qquad (6.120)$$

▯

PROOF

The proofs follow the paper of [19].

▯

REMARK 6.19 For the fix-point problem, we can stabilize our scheme by the linearization of the last solutions. With more iterations, the stabilization is obtained. Here is also a balance between the time partitions and the iterations important to save computational time.

▯

In the next section our numerical examples are based on the magnetization of isotropic and anisotropic materials.

The isotropic material on the one hand is hard to control, where the anisotropic material has special directions to follow by the discretization.

6.5.4.4 Test Example of the LLG Equation with Pure Isotropic Case

We start with a first example to use the stabile first-order splitting as a prestep method. In the test example for the pure isotropic case, we have a situation, without an influence by an outer magnetic field:

$$u_t = u \times \Delta u - \lambda u \times (u \times \Delta u), \tag{6.121}$$

where $\lambda = 1$, $\Omega = [\text{-}0.5, 0.5]^2$.

We concentrate on different initial values that complicate the computations
1. Case

$$u_0 = \begin{cases} (2xA; A^2 - |x|^2)/(A^2 + |x|^2) & \text{for } |x| < 0.5 \\ (0;0;-1) & \text{for } |x| > 0.5 \end{cases} \tag{6.122}$$

2. Case

$$u_0 = \begin{cases} (2xA; A^2 - |x|^2)/(A^2 + |x|^2) & \text{for } |x| < 0.05 \\ (0;0;-1) & \text{for } |x| > 0.05 \end{cases} \tag{6.123}$$

3. Case

$$u_0 = [2xA; A^2 - |x|^2]/(A^2 + |x|^2) \text{ for } x \in \Omega \tag{6.124}$$

where $A = (1 - 2|x|)^4/4$.
The convergence-results are given in Tables 6.46 and 6.47:

TABLE 6.46: Relative errors between $l = 4$ and $l = 6$.

x	y	L_2 (relative error)	L_∞ (relative error)
0	0	0.487524	1.999790
−1	0	0.201978	0.426290
0	−1	0.199584	0.426297
1	0	0.198766	0.426282
0	1	0.201194	0.426287

The results show the switching of the singular point in more unstable situations as the nonsingular points.
The computational results of the singular point are presented in Figure 6.26.
The influence of the external field is presented in Figure 6.27.
In a next experiment one can present a more stable configuration with an external field.

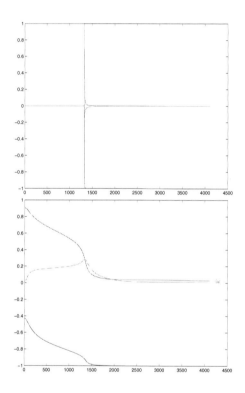

FIGURE 6.26: Left: Magnetization on the point $(0,0)$, right: magnetization on the point $(1,0)$ with the coordinates: red $= x$, green $= y$, blue $= z$.

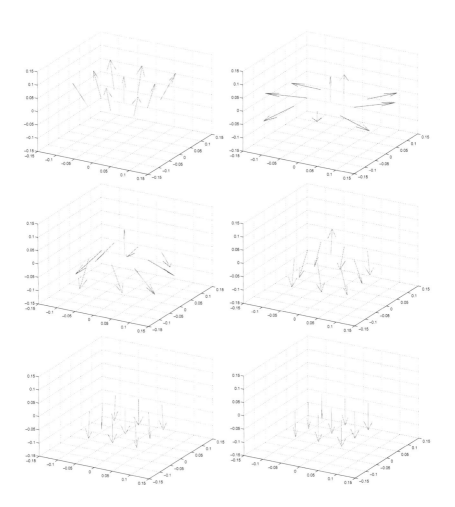

FIGURE 6.27:   Figures top-down and left-right present time-sequence of a magnetic layer in a pure isotropic case.

TABLE 6.47: Relative errors between $l = 5$ and $l = 6$.

x	y	L_2 (relative error)	L_∞ (relative error)
0	0	0.662921	1.999939
−1	0	0.191115	0.453568
0	−1	0.194658	0.468610
1	0	0.198813	0.493772
0	1	0.195291	0.479565

6.5.4.5 Test example of the LLG Equation with External Field $H_{eff}(t)$

The following problem is discussed as magnetization based with an external magnetic field.

$$u_t = u \times (\Delta u + H_{eff}) - \lambda u \times (u \times (\Delta u + H_{eff})), \qquad (6.125)$$

where $\lambda = 1$, $\Omega = [\text{-}0.5, 0.5]^2$, and $\alpha = 1$, $H_{ext} = (1, 0, 40)$ (e.g., with a strong right-hand side influence).

We concentrate on different initial values that are given as

$$u_0 = \begin{cases} (2xA; A^2 - |x|^2)/(A^2 + |x|^2) & \text{for } |x| < 0.5 \\ (0; 0; -1) & \text{for } |x| > 0.5 \end{cases}. \qquad (6.126)$$

The spatial discretization is done with 9×9 points in the domain, and the time discretization is based on a time-step $\Delta t = 1/640$.

In Figure 6.28 we present the influence of the external field.

The interaction between the particles is very strong with the help of the H_{ext}. The internal vector has a rapid change, due to the symmetry break (i.e., with H_{ext}). In the next time-steps the magnetization vectors are oriented to the external field (i.e., to the top of the material).

REMARK 6.20

We present the combination with isotropic and anisotropic magnetization behavior. Stable discretization methods based on penalizing methods are described, further splitting methods could decouple the complicated equations. In numerical experiments we verify our theoretical results. In the future the nonsmooth treatment of magnetic fields with comparison of experimental dates is proposed.

□

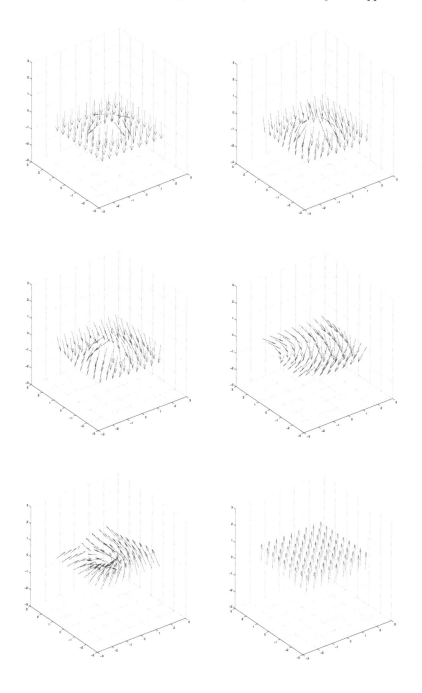

FIGURE 6.28:   Figures top-down and left-right present time-sequence of a magnetic layer influenced by an external magnetic field $H_ext = (1, 0, 40)$.

Chapter 7

Summary and Perspectives

The monograph presented splitting analysis for evolution equations. In several chapters, both time and space scales are treated and balanced to achieve accurate and stable methods.

The numerical analysis for the iterative operator-splitting methods is presented to quasi-linear, stiff, spatiotemporal problems. Further, we discuss the applications to hyperbolic equations, with respect to systems of wave equations.

The following contributions are treated:

- Generalization of consistency and stability results to nonlinear, stiff and spatial decomposed splitting problems

- Acceleration of the computations by decoupling into simpler problems

- Efficiency of the decomposed methods

- Theory based on semigroup analysis which is well understood and applicable to the splitting methods

- Applications in computational sciences (e.g., flow problems, elastic wave propagation, heat transfer).

Based on the iterative splitting methods, the important fix-point problem, with respect to the initial solutions or starting solutions, is important.

So the following items should be further discussed and developed in algorithms:

- Improved starting solution for the first iterative step by solving the fix-point problem accurate (e.g., embedded Newton-solver)

- Parallelization of the iterative methods on the operator level

- Decoupling algorithms with respect to the spectrum of the operators and atomization the decomposition process

- Simulation of real-life problems with improved parallel splitting methods

The perspective for the decomposition methods is combing the efficiency decoupling and generalization due to nonlinear and boundary problems. The decomposition methods are adaptive methods that respect the physical conservation of the equation and accelerate the solving process of a complicated system of coupled evolution equations.

Chapter 8

Notation

8.1 List of Abbreviations

- FOP : First-order operator splitting

- FOPSWR : First-order operator splitting with Schwarz waveform-relaxation methods

- DD : Domain decomposition methods

- ADI : Alternating direction implicit methods

- LOD : Local one-dimensional methods (e.g., local one-dimensional splitting methods)

- ILU : Incomplete LU method

- PDE : Partial differential equation

- ODE : Ordinary differential equation

- R^3T : Radioactive-Reaction-Retardation-Transport software toolbox, done with the software package UG (unstructured grids)

- d^3f : Distributed-Density-Driven-Flow software toolbox, done with the software package UG (unstructured grids)

- $WIAS - HiTNIHS$: WIAS High-Temperature Numerical Induction Heating Simulator (software package for heat and radiation equations, designed for crystal-growth applications)

- $OPERA - SPLITT$: Program software package including splitting methods

- PVT : Physical vapor transport

- SiC : Silicon-carbide

- CFL condition : Courant-Friedrichs-Levy condition

- DDL : Dynamical link library

- OFELI : Object Finite Element Library

- COM : Component object model

8.2   Symbols

- λ - eigenvalue

- $\rho(A)$ - spectral radius of A

- e_i - eigenvector

- $\sigma(A)$ - spectrum of A

- $c_t = \frac{\partial c}{\partial t}$ - first-order time derivative of c

- $c_{tt} = \frac{\partial^2 c}{\partial t^2}$ - second-order time derivative of c

- $\tau = \tau_n = t^{n+1} - t^n$ - time-step

- c^n - approximated solution of c at time t^n

- $\partial_t^+ c = \frac{c^{n+1} - c^n}{\tau_n}$ - forward finite difference of c in time

- $\partial_t^- c = \frac{c^n - c^{n-1}}{\tau_n}$ - backward finite difference of c in time

- $\partial_t^2 c = \partial_t^+ \partial_t^- c$ - second-order finite difference of c in time

- ∇c - gradient of c

- $\Delta c(x,t)$ - Laplace operator of c

- $\mathbf{n}$ - outer normal vector

- $\partial_x^+ c$ - forward finite difference of c in space dimension x

- $\partial_x^- c$ - backward finite difference of c in space dimension x

- $\partial_y^+ c$ - forward finite difference of c in space dimension y

- $\partial_y^- c$ - backward finite difference of c in space dimension y

- $e_i(t) := c(t) - c_i(t)$ - local error function with approximated solution $c_i(t)$

- err_{local} - local splitting error

- err_{global} - global splitting error

- $arg(z)$ - argument of $z \in \mathbb{C}$ (angle between real and imaginary axes)

- J_λ - nonlinear resolvent

- A_λ - Yoshida approximation

- J_A - Jacobian matrix

- $[A, B] = AB - BA$ - commutator of operators A and B

8.3   General Notations

$D(B)$	Domain of B
$\mathbf{X}, \mathbf{X}_E$	Banach spaces
$\mathbf{X}^n = \Pi_{i=1}^n \mathbf{X}_i$	Product space of $\mathbf{X}$
$W^{m,p}(\Omega)$	Sobolev space consists of all locally summable functions $u : \Omega \to \mathbb{R}$ such that for each multi-index α with $\lvert\alpha\rvert \le m$, $D^\alpha u$ exists in the weak sense and belongs to $L^p(\Omega)$
$\partial\Omega$	Boundary of Ω
$\{S(t)\}$	Semi-group (see Section 2.2.5)
$\mathcal{L}(X) = L(X, X)$	Operator space of $\mathbf{X}$, see Theorem 3.1
Ω_h	Discretized domain Ω with the underlying grid-step h.
H^m	Sobolev space $W^{m,2}$
$H_0^1(\Omega)$	The closure of $C_c^\infty(\Omega)$ in the Sobolev space $W^{1,2}$, see Section 2.2
$\lVert \cdot \rVert_{L^2}$	L^2-norm
$\lVert \cdot \rVert_{H^m}$	H^m-norm
$\lVert \cdot \rVert$	Maximum norm, if not mentioned otherwise
$\lVert \cdot \rVert_{\mathbf{X}}$	Norm with respect to Banach-space $\mathbf{X}$
(x, y)	Scalar product of x and y in a Hilbert space
$\mathcal{O}(\tau)$	Landau symbol (e.g., first-order in time with time step τ)

Appendix A

Software Tools

A.1 Software Package UG (P. Bastian et. al. [18])

In this section we introduce the software package r^3t. The real-life problems of the waste disposal scenarios are integrated in this package and all numerical calculations were performed with it.

The software package was developed for the task of the project to simulate transport and reaction of radioactive pollutants in groundwater. The specifications for the software package were flexible inputs and outputs as well as the use of large time-steps and coarse grids to achieve the claimed long-time calculations.

The concept of the **UG** software toolbox and the different software packages **UG**, d^3f, r^3t, and **GRAPE** are to the fore.

For satisfying the tasks, we built on the preliminary at the institute, developed **UG** software toolbox. We use the name software toolbox to reveal that the software tools can be provided for other software packages. The software will not be concluded in the sense of a software package, but will be extendable. The development consisted of compiling a flexible input interface. Further a coupling concept was developed that uses data from other software packages. Particularly, the implementation of efficient discretizations and error estimators was realized to achieve the claimed large time-steps and coarse grids for the numerical calculations.

The main topics of this chapter consist of the rough structuring of the software packages, the description of the **UG** software toolbox, the description of the d^3f software package, the description of the r^3t software package, as well as their concepts.

A.1.1 Rough Structuring of the Software Packages

Software Toolbox **UG**: The **UG** software toolbox manages unstructured, locally refined multigrid hierarchies for two- and three-dimensional geometries. The software toolbox provides the software structures for the further modules. It has a variety of usable libraries for discretizations and solutions of systems of partial differential equations, as well as interactive graphics to display results of simulations. Therefore, it serves as a basic module for the development and programming of the **r³t** and **d³f** software packages.

Software Package **d³f**: This package was developed to solve densely floated fluxes of groundwater. It consists of a flux and a transport equation that are nonlinear and coupled. The simulation calculations with software packages can treat steady as well as unsteady fluxes. The software package has a flexible input setting and yields velocity data for the transport calculations of the **r³t** software package.

Software Package **r³t**: The software package was developed for solving transport-reaction equations of various species in flowing groundwater through porous media. For the solutions of convection-dominant equations, an improvement of the discretization of equations was developed. The goal was to achieve large time-steps and coarse grids to gain the model times in a claimed calculation time. The error estimators and solvers were adapted to these finite volume discretizations. The flexible input of parameters for the model equation as well as the output of the data of solution and grid are further tasks that the software has to fulfill.

Software Package **GRAPE**: The software package is a wide-ranging visualization software that is particularly suited for visualizations of solutions on unstructured grids with various presentation methods. The results of the **r³t** module are visualized using this package. Wide-ranging visualization possibilities of **GRAPE**, see [111], could be used to gain an improved comprehension of the simulation results.

A.1.2   UG Software Toolbox

The basis of all subsequently described software packages is the **UG** software toolbox. It is derived from the words "**U**nstructured**G**rids" and is a software toolbox for solving a system of linear or nonlinear partial differential equations.

The concept of the **UG** software toolbox is the management of an unstructured, locally refined multigrid hierarchy for two- and three-dimensional grids.

The software package was developed in the beginning of the 1990s, see [18], using the adaptivity of grids, multigrid methods, and parallelization of the methods to solve systems of partial differential equations. Because of effective and parallel algorithms of the software package, one has the opportunity to solve wide-ranging model problems.

Several works accrued in the field of numerical modeling and numerical simulation of linear and nonlinear partial differential equations, mechanics of elasticity, densely floated fluxes, transport of pollutants through porous media, as well as mechanics of fluxes. Some examples can be found in [17], [18], and [127].

Subsequently, we introduce the application of the **UG** software toolbox.

For the application of the software toolbox **UG** in a software package, as e.g. **r³t**, we distinguish between two modules when structuring the emerging software package.

ug: This module is the part of the software package which is independent of the application. It contains software tools as the interpreter of commands, the event management, the graphics, the numerics, the local grid hierarchy, the description of areas, the device connection, as well as the storage management. The lowest level contains the parallelization of the program using a graph-based parallelization concept. These tools can be applied for the further part of the application-oriented area.

r3t: In this module the model problem with the related equation is prepared for the application. For the implementation, the problem class library and the application were developed. The module depends on the application in contrast to the **ug** module that is independent of the application.

The following classification was constituted: Discretizations depending on the problem as well as the according error indicators, the sources and sinks, and the analytical solutions, that were developed especially for these systems of differential equations, are implemented in the program class library.

For the discretization, the finite volume method of higher order was used, see Appendix B.3. As error indicators for the convection-diffusion-dispersion equation, the *a posteriori* error indicators were used, see [163].

The applications contain the initial and boundary conditions as well as the coefficient functions for the equation. Further implemented are the flexible input of equation parameters and several time-independent source terms. In addition, script files are included in the application. These are loaded by the **UG** command interpreter and control the sequence of calculations. Especially the scanner and parser module were programmed and implemented into the application. At a cue by the program are set: the number of equations, the phases with transport and reaction parameters, the used sources, the boundary conditions, the area parameter, as well as the hydrogeological parameters. All files used for the input of velocity and geometry data as well as the output of solution and grid data can be found in the application.

A.1.3 UG Concept

Essentially two concepts of the **UG** software toolbox were also used for the software package **r³t**. One concept is the unstructured grid concept that allows us to manage relatively complex geometries with adaptive operations to make the hierarchical methods applicable.

A further concept is the concept of sparse matrices. It allows the block by block savings of wide-ranging systems of partial differential equations to achieve an effective storage and velocity concept.

These two concepts are for determining the efficiency of the **r³t** software package and are described subsequently.

We first describe the concept of unstructured grids. The basis for the **UG** software toolbox consists of the creation and modification of unstructured grids. Using unstructured grids, complex geometries can be approximated.

An efficient structure of the grid data is necessary to produce grids and perform local refinements as well as coarsenings. The geometrical data structure administers geometrical objects, as, for example, elements, edges, and nodes. The algebraic data structure administers matrices and vectors.

The grids consist of triangles and quadrangles in two-dimensions, or tetrahedrons, pyramids, hexahedrons, and prisms. The structure of the grid data is constructed hierarchically and the elements can be locally refined and coarsed.

The application of multigrid methods with varying smoothing processes is possible because of the locally refineable areas.

The solutions on higher grid levels are only considered on the refined elements. Hence, local grids are treated with all refined elements and some copied elements. One obtains an efficient saving and a high efficiency in processing.

A further important concept for our efficient calculations with the r^3t software package is the saving of sparse matrices that are used in the **UG** software toolbox.

Sparse matrices occur when grids are saved. Based on the local discretizations, only the neighbors are used for the calculations, such that only some entries appear outside of the diagonals. Sparse matrices also occur at the application of weakly coupled systems of partial differential equations. The connections between the individual equations are saved as entries outside of the diagonal. Due to the weak coupling only a few entries appear. To avoid the ineffective saving of fully populated matrices, the saving of sparse matrices was developed.

An efficient storage concept, the sparse matrix storage method [161] was used. The idea of the method is based on the saving of sparse matrices in a block-matrix graph, in which blocks are indicated as local arrays over a set of compact row-orientated patterns, see [161]. This combines the flexibility of graph structures with the higher efficiency based on higher data densities. The compact inner pattern allows us to identify entries, with the further advantages of saving and calculation times that can be obtained. The compact storage technique is row or column oriented, preliminary individual matrix entries can be replaced by block entries that represent a system of partial differential equations. The entry in the diagonal of the block matrix is the coupling within the equations, the entry outside the diagonal of the block matrix is the coupling between neighboring nodes, if a space term is used. The saving as vector-matrix graph is considered in Figure A.1.

Hence, the claimed conditions to solve wide-ranging systems of partial differential equations on complex geometries were satisfied.

Matrix Objects

● Vector Objects

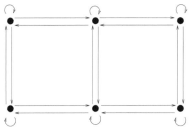

FIGURE A.1: Vector-matrix graph for efficient saving.

A.1.4 Software Package d³f

The previous project about the **r³t** software package was posed by the GRS in Braunschweig. It was developed to simulate densely floated fluxes of groundwater. It contains an implementation of the catalogs for problem classes and applications. In the problem class, error estimators, solvers, and discretizations for the equations were implemented. The catalog for applications contains several applications. Particularly implemented are special boundary and initial conditions, special prefactors for the equations, as well as sources and sinks. The initialization of the equation parameters, as well as boundary and initial conditions, and the parameters for sources and sinks are read in from files at the start time. Therefore, flexible calculations can be made in two- and three-dimensional model examples.

A.1.5 Equations in d³f

Two coupled nonlinear time-dependent partial differential equations were implemented in the software package **d³f**. These equations are subsequently considered.

The equation for the flux is given as

$$\partial_t(\phi\rho) + \nabla \cdot (\rho\mathbf{v}) = Q, \tag{A.1}$$

where ρ denotes the density of the fluid, $\mathbf{v}$ calculates the velocity using Darcy's law, ϕ is the effective porosity, and Q stands for the source term.

We use Darcy's law to derive the velocity of the fluid through the porous media. Darcy's law is a linear relationship between the discharge rate through a porous medium, the viscosity of the fluid, and the pressure drop over a given distance:

$$Q = \frac{-\kappa A}{\mu} \frac{P_b - P_a}{L}. \tag{A.2}$$

The total discharge, Q (units of volume per time), is equal to the product of the permeability (κ units of area) of the medium, the cross-sectional area A to flow, and the difference of pressure drops P_b and P_a, all divided by the dynamic viscosity μ, and the length L of the pressure drop. Dividing both sides of the equation by the area, we obtain the general notation:

$$q = \frac{-\kappa}{\mu} \nabla P, \qquad (A.3)$$

where q is the flux and ∇P is the pressure gradient vector. This value of flux is often called the *Darcy flux*.

The pore velocity v is related to the Darcy flux q by the porosity ϕ and is given as

$$v = \frac{q}{\phi}. \qquad (A.4)$$

The pore velocity is used as the velocity of a conservative tracer and is used as our velocity for the simulations given by physical experiments.

The transport equation is given as

$$\partial_t(\phi \rho c) + \nabla \cdot (\rho \mathbf{v} c - \rho D \nabla c) = Q', \qquad (A.5)$$

where c is the concentration of the solute, D denotes the diffusion-dispersion tensor, and Q' is the source term.

The first equation (A.1) calculates the flux through porous media using Darcy's law as the continuity equation and gives the density of the fluid. The second equation is a transport equation for water in dissolved salt. It is a time-dependent convection-diffusion-dispersion equation, where the convective factor is calculated using Darcy's law. Depending on the distinction of density, vortical fluxes can occur. Based on the small diffusions in water, one obtains a convection-dominated equation, for this equation adapted solver and discretization methods were developed, see [127].

A.1.6 Structure of d³f

For the input of equation and control parameters a preprocessor was developed. The preprocessor reads the wide-ranging input parameters before the calculations start. The essential module and centerpiece make up the processor that was developed on the basis of **UG** and adapted with further problem-specific classes for the used equations. Applications that contain the parameter files and control files for the program run, are filed in a further library. For data to be circulated, a postprocessor was developed that presents an interface and saves the data of the solutions in files. As a result, other programs, especially the visualization software **GRAPE**, can read the data and reprocess them.

A.2 Software Package r³t

In this section we roughly present the software package **r³t**, the name is derived from "**R**adionuclide, **R**eaction, **R**etardation, and **T**ransport" (**r³t**). The software package discretizes and solves systems of convection-diffusion-dispersion-reaction equations with equilibrium sorption and kinetic sorption.

A.2.1 Equation in r³t

The equation for the equilibrium sorption is given by

$$
\phi \, \partial_t \, R_i \, c_i + \nabla \cdot (\mathbf{v}c_i - D\nabla c_i)
$$
$$
= -\phi \, R_i \, \lambda_i c_i + \sum_{k=k(i)} \phi \, R_k \, \lambda_k c_k + \tilde{Q}_i, \qquad (A.6)
$$
$$
c_{e(i)} = \sum_i c_i,
$$
$$
R_i = 1 + \frac{(1-\phi)}{\phi} \rho \, K(c_{e(i)}),
$$
$$
\text{with} \quad i = 1, \ldots, M,
$$

where c_i denotes the i-th concentration, R_i the i-th retardation factor, λ_i the i-th decay factor, and $\mathbf{v}$ the velocity that is either calculated in the **d³f** software package or preset. Q_i denotes th i-th source term, ϕ the porosity, $c_{e(i)}$ the sum of all isotope concentrations referring to element e, K is the function of the isotherms, see Appendix B.3.

A.2.2 Task of r³t

The software package was commissioned by the GRS for prognoses in the area of radioactive repositories in salt deposits.

The following properties were claimed for the development of the software package **r³t**:

- Independence of the transport equation that has to be solved, shall be achieved. This can be obtained by applying a two- or three-dimensional version based on the flexible concept.

- A simple adaptability on shell-level shall be possible. A script file with preferences for the parameters of discretization and solver, as well as a graphical visualization shall be used. A suitable definition of interfaces for the output of data for reprocessing by a graphic software as, for example, **GRAPE**, confer [111], will be done.

- The programming of self-contained modules will be developed, the modules will outwardly be structured with accurately defined interfaces. Therefore, several programmers can be involved in the development.

- The differential equations will be applied as scalars and also as systems. The coupling of at least four phases will be possible.

- All procedures will be possible for a simultaneous computer to allow wide-ranging calculations.

The structure of the software package $\mathbf{r^3t}$ is based on the preliminary work of the developed software toolbox **UG**, see Section A.1.2, and was advanced in this context. The presented tasks were realized using the following concepts.

A.2.3 Conception of r^3t

The software conception of $\mathbf{r^3t}$ is given with the following approaches.

While developing our software package $\mathbf{r^3t}$, the philosophy and conception of the **UG** development further evolved. User-independent numerical algorithms were programmed and provided for special applications. During this development of software package $\mathbf{r^3t}$, further numerical algorithms concerning discretizations, solvers, and error indicators were implemented. The used application-independent structures, the *numerical procedures*, were continued, confer [128].

The essential renewals of this development phase consisted of connecting the existing software packages and combining the corresponding tasks of the respective programs. Thus the software package $\mathbf{d^3f}$, that is used to calculate the flux of the transport equation, [72] and [127], was coupled with the software package $\mathbf{r^3t}$. A coordinate-oriented and grid-independent storage of the velocity data was developed. A further connection to the visualization software **GRAPE**, confer [176], could be used to visualize the results.

A flexible input was developed for the wide-ranging parameters that are available in the form of input files. Hence, the wide-ranging test calculations are possible which are necessary to achieve qualified statements. This further flexibility is adequate for the concept of modular programming. Individual parts of the program, as the preprocessor and the corresponding discretization and solution methods, can be developed independently. A further flexibility is obtained from the data concept that enables the input and output of files using interfaces, such that the individual software packages were coupled.

The application of software package $\mathbf{r^3t}$ with used connections to other software packages is subsequently described.

A.2.4 Application of r^3t

The application of $\mathbf{r^3t}$ is outlined in Figure A.2.

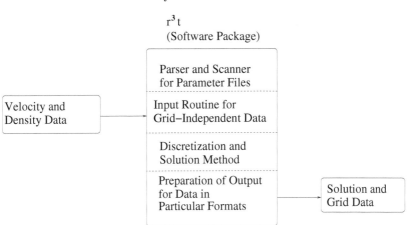

FIGURE A.2: Application of $\mathbf{r^3t}$.

The approach of software package $\mathbf{r^3t}$ illustrated in Figure A.2 is explained subsequently.

The preparations to the point where the model problems are calculated consist of the input of equation parameters and control files. Particularly, the number of equations and phases as well as the parameters for convection-diffusion-dispersion-reaction equations can be ascertained. These are read in during the start of the initialization phase, as, for example, the data of geometry and velocity that are grid-independently stored in files by software package $\mathbf{d^3f}$.

The actual calculation phase of the software package proceeds after the initialization. The problem-specific discretization as well as the solvers for linear equation systems are used. The error indicators are used for adaptivity to realize the effective calculations in time of equations with local grid refinements. Especially velocity and density data are used for adaptive calculations. The individual subequations that were treated using various discretizations, see Appendix B.3, are coupled afterwards with the operator-splitting methods.

The results of the calculation steps are read out into a particular file format in the output phase and are processed afterwards by the postprocessor **GRAPE**. During the preparations the total concentration is calculated by using the explicit correlation of mobile and adsorbed concentration, confer Equation (1.1), for the case of equilibrium-sorption. The associated grid data belong to the output data of solutions and describe the geometric proportions of the data.

Hence, the application of the $\mathbf{r^3t}$ software package was integrated in ambit of further software packages that realize preprocessing or postprocessing.

We subsequently describe a coupling concept to achieve a flexible interchange of files between the involved software packages.

A.2.5   Coupling Concept of r³t

The concept of coupling between the individual software packages is now developed based on interchanges of files at predefined interfaces. This connection is sufficient for parameters that are only insignificantly changed during the time-dependent course and have no backcoupling on other programs.

The problem-dependent parameters that are constant during the calculation time can be read in at the beginning over files, as, for example, geology of rocks in the area, exchange parameters of radioactive species, as well as determination of sources and sinks.

Further inputs can vary in the course of the simulation time, as, for example, velocity or density data. These are read in during the duration in the specific time points. The data are coupled single sided to the **r³t** software package that is, the results of **r³t** are not used for the calculation. The reason is the modeling, where the radioactive contaminants have no backcoupling on the flow field. The backcoupling on software packages **d³f** therefore can be omitted.

Completed calculations are saved in files and for further utilization read in from the postprocessor.

A strong coupling between the two programs was not pursued further. This is only advisable for a coupling between flux and contaminants, then transport and flux equation are solved together.

Using software package **r³t**, the subsequent model examples can be realized and evaluated.

A.3   Solving Partial Differential Equations Using FIDOS (authors: J. Geiser, L. Noack)

This section describes the treated PDEs, the used methods, and the usage of the program package FIDOS 1.0 (Finite Difference and Operator-Splitting Methods).

The first version, FIDOS 1.0, covers only computations that already have been published and used in this monograph. In the following subsections we discuss the treatment of the methods and the assembling and the programming structure.

A.3.1   Treated PDEs

The PDEs treated in this paper consist of the form

$$D_t u = D(u)u, \tag{A.7}$$

where the derivative in time is of first or second order,

$$D_t u = \partial_t u \text{ or } D_t u = \partial_{tt} u, \tag{A.8}$$

and the spatial operator $D(u)$ can consist of derivatives in one to three spatial dimensions depending on the PDE:

$$D(u)u = D_x(u)u + D_y(u)u + D_z(u)u. \tag{A.9}$$

The spatial operators can consist of linear terms, for example,

$$D_x(u) = A\partial_x u \text{ or } D_x(u) = A\partial_{xx} u, \tag{A.10}$$

and can contain nonlinear terms, as well,

$$D_x(u) = Au\partial_x u, \tag{A.11}$$

where A (or B, C, respectively) is a constant.

The types of PDEs that can be examined using FIDOS, therefore varies from simple wave and heat equations to nonlinear momentum equations. The package can easily be extended to other kinds of PDEs and more methods. The PDEs of the first version of the program package are listed below.

A.3.1.1 Wave Equation

The wave equation for three dimensions is given by

$$\partial_{tt} u = A\partial_{xx} u + B\partial_{yy} u + C\partial_{zz} u, \text{ with} \tag{A.12}$$

$$u(x, y, z, 0) = \sin\left(\frac{1}{\sqrt{A}}\pi x\right) \sin\left(\frac{1}{\sqrt{B}}\pi y\right) \sin\left(\frac{1}{\sqrt{C}}\pi z\right), \tag{A.13}$$

$$\partial_t u(x, y, z, 0) = 0. \tag{A.14}$$

The analytical solution in $\Omega = [0, 1]^3$ is given by

$$u(x, y, z, t) = \sin\left(\frac{1}{\sqrt{A}}\pi x\right) \sin\left(\frac{1}{\sqrt{B}}\pi y\right) \sin\left(\frac{1}{\sqrt{C}}\pi z\right) \cos\left(\sqrt{3}\pi t\right). \tag{A.15}$$

The one- and two-dimensional cases follow analogously, whereby the constant $\sqrt{3} = \sqrt{d}$ has to be changed according to the dimension d.

The wave equation can be treated for a nonstiff case (e.g., $A = B = C = 1$), as well as for stiff PDEs (e.g., $A = 0.01, B = 1, C = 100$). See [100] and [104] for further explanations.

A.3.1.2 Viscous Burgers Equation

The viscous Burgers equation is given as

$$\partial_t u = -u\partial_x u - u\partial_y u + \mu(\partial_{xx} u + \partial_{yy} u), \tag{A.16}$$

where the first two terms of the right-hand side together are the nonlinear operator $A(u)u$, whereas the last term can be regarded as a linear operator Bu.

The analytical solution of this PDE is

$$u(x, y, t) = \left(1 + e^{(x+y-t)/2\mu}\right)^{-1}. \tag{A.17}$$

A.3.1.3 Mixed Convection-Diffusion and Burgers Equation

The mixed convection-diffusion and Burgers equation is

$$\partial_t u = -\frac{1}{2}(u\partial_x u + u\partial_y u + \partial_x u + \partial_y u) + \mu(\partial_{xx}u + \partial_{yy}u) + f(x, y, t), \tag{A.18}$$

here the analytical solution is chosen to be

$$u(x, y, t) = \left(1 + e^{(x+y-t)/2\mu}\right)^{-1} + e^{(x+y-t)/2\mu}. \tag{A.19}$$

The function f is calculated accordingly. The equation is again split into a nonlinear part and a linear part to obtain the operators $A(u)u$ and Bu.

A.3.1.4 Momentum Equation

The momentum equation is given as

$$\partial_t \mathbf{u} = -\mathbf{u} \cdot \nabla \mathbf{u} + 2\mu\nabla(\mathbf{u} \cdot \mathbf{u} + \mathbf{u} \cdot \mathbf{v} + 1/3\nabla\mathbf{u}) + \mathbf{f}(x, y, t), \tag{A.20}$$

with analytical solution $\mathbf{u} = (u_1, u_2)^T$ and

$$u_1(x, y, t) = u_2(x, y, t) = \left(1 + e^{(x+y-t)/2\mu}\right)^{-1} + e^{(x+y-t)/2\mu}. \tag{A.21}$$

The function f is again calculated accordingly.

The one-dimensional formulation follows analogously, see [101].

A.3.1.5 Diffusion Equation

Being a diffusion equation, the two-dimensional heat equation is treated for an example. The used methods can easily be applied to any other diffusion equations:

$$\partial_t u = A\partial_{xx}u + B\partial_{yy}u, \tag{A.22}$$

with analytical solution

$$u(x, y, t) = e^{-(A+B)\pi^2 t}\sin(\pi x)\sin(\pi y). \tag{A.23}$$

A.3.2 Methods

In general, all used methods, independent of their explicit or implicit character and independent of the number of needed substeps, satisfy

$$u^{n+1}(x, y, z, t) = METHOD(u^n(x, y, z, t), u^{n-1}(x, y, z, t)) \qquad (A.24)$$

with given initial condition

$$u^0(x, y, z) \text{ for } t = 0, \qquad (A.25)$$

and if needed

$$u^1(x, y, z) \text{ for } t = 0, \qquad (A.26)$$

if the space consists of three dimensions, analogous formulations hold for less dimensions. In the following, the number of dimensions is always set to be two, if not mentioned otherwise. All methods can be applied to one or three spatial dimensions, too.

A.3.2.1 ADI Method

The ADI method is given subsequently for two-dimensional equations with first-order derivatives in time.

The Crank-Nicholson method,

$$\left(L_1 L_2 + \frac{\Delta t}{2} \left(L_2 A \partial_{xx} + L_1 B \partial_{yy} \right) \right) u^{n+1}$$
$$= \left(L_1 L_2 - \frac{\Delta t}{2} \left(L_2 A \partial_{xx} + L_1 B \partial_{yy} \right) \right) u^n \qquad (A.27)$$

with some operators L_1, L_2, is split into two substeps, yielding the ADI method:

$$(L_1 + \frac{\Delta t}{2} A \partial_{xx}) u^{n+1,1} = (L_1 - \frac{\Delta t}{2} A \partial_{xx})(L_2 - \frac{\Delta t}{2} B \partial_{yy}) u^n, \quad (A.28)$$
$$(L_2 + \frac{\Delta t}{2} B \partial_{yy}) u^{n+1} = u^{n+1,1}. \qquad (A.29)$$

Depending on the operators L_1, L_2, the ADI method is a second- or fourth-order method. The first case is obtained with $L_1 = L_2 = 1$, the second one with $L_1 = 1 + \Delta x^2/12 \, \partial_{xx}$ and $L_2 = 1 + \Delta y^2/12 \, \partial_{yy}$.

A.3.2.2 LOD Method

The LOD method below is derived for partial differential equations with two spatial dimensions, linear operators $D_x u$, $D_y u$, and first- and second-order derivatives in time:

$$\partial_t u = (D_x + D_y) u,$$
$$\partial_{tt} u = (D_x + D_y) u.$$

Using the finite difference schemes

$$\partial_t u^n \approx \frac{u^{n+1} - u^n}{\Delta t} \quad \text{and} \tag{A.30}$$

$$\partial_{tt} u^n \approx \frac{u^{n+1} - 2u^n + u^{n-1}}{\Delta t^2}, \tag{A.31}$$

the LOD method is derived from the explicit Euler method,

$$u^{n+1} - u^n = \Delta t (D_x + D_y) u^n, \tag{A.32}$$

$$u^{n+1} - 2u^n + u^{n-1} = \Delta t^2 (D_x + D_y) u^n. \tag{A.33}$$

The LOD method for PDEs with first-order derivatives in time is now given by

$$u^{n+1,0} - u^n = \Delta t (D_x + D_y) u^n, \tag{A.34}$$

$$u^{n+1,1} - u^{n+1,0} = \Delta t \eta D_x (u^{n+1,1} - u^n), \tag{A.35}$$

$$u^{n+1} - u^{n+1,1} = \Delta t \eta D_y (u^{n+1} - u^n), \tag{A.36}$$

and the LOD method for second-order derivatives is

$$u^{n+1,0} - 2u^n + u^{n-1} = \Delta t^2 (D_x + D_y) u^n, \tag{A.37}$$

$$u^{n+1,1} - u^{n+1,0} = \Delta t^2 \eta D_x (u^{n+1,1} - 2u^n + u^{n-1}), \tag{A.38}$$

$$u^{n+1} - u^{n+1,1} = \Delta t^2 \eta D_y (u^{n+1} - 2u^n + u^{n-1}). \tag{A.39}$$

A.3.3 Iterative Operator-Splitting Methods

The iterative methods repeat the solving steps until a given break criterion is achieved. Such a criterion can be the number of iteration steps (as is used in FIDOS) or an error tolerance. Then the general scheme for an iterative method is

$$u^{n+1,i} = METHOD(u^{n+1,i-1}), \quad i = 1 \dots m \tag{A.40}$$

$$u^{n+1} = u^{n+1,m}. \tag{A.41}$$

There are different possibilities to choose the initial value $u^{n+1,0}$, and all are integrated in FIDOS:

$$\begin{aligned}
&(1)\ u^{n+1,0} = u(x, y, t^{n+1}), \\
&(2)\ u^{n+1,0} = 0, \\
&(3)\ u^{n+1,0} = u^n.
\end{aligned} \tag{A.42}$$

If two initial values are needed, they are set to be equal:

$$u^{n+1,1} = u^{n+1,0}. \tag{A.43}$$

A.3.3.1 Standard IOS Method

The standard iterative operator-splitting method is given with the following scheme, where the partial derivatives are exchanged by usage of (A.30).

The IOS method for equations with first-order derivatives in time is given as

$$\partial_t u^{n+1,i} = D_x(u^{n+1,i-1})u^{n+1,i} + D_y(u^{n+1,i-1})u^{n+1,i-1}, \quad (A.44)$$
$$\partial_t u^{n+1,i+1} = D_x(u^{n+1,i-1})u^{n+1,i} + D_y(u^{n+1,i-1})u^{n+1,i+1} \quad (A.45)$$

for $i = 1, 3, \ldots, 2m + 1$.

A.3.3.2 Coupled η-IOS Method

For the wave equation, the second-order derivatives in time are replaced using the finite difference scheme (A.31). This method is explicit, but the coupling of the current time-step t^{n+1} with the two previous time-steps using the η-method is much more effective. The η-method uses the approximation

$$u^n \approx \eta u^{n+1} + (1 - 2\eta)u^n + \eta u^{n-1}, \quad (A.46)$$

for $\eta \in [0, 1/2]$. Equality holds for $\eta = 0$.

Then the scheme of the coupled η-IOS method is given by

$$u^{n+1,i} - 2u^n + u^{n-1} = \Delta t^2 A(\eta u^{n+1,i} + (1 - 2\eta)u^n + \eta u^{n-1}) \quad (A.47)$$
$$+ \Delta t^2 B(\eta u^{n+1,i-1} + (1 - 2\eta)u^n + \eta u^{n-1}),$$
$$u^{n+1,i+1} - 2u^n + u^{n-1} = \Delta t^2 A(\eta u^{n+1,i} + (1 - 2\eta)u^n + \eta u^{n-1}) \quad (A.48)$$
$$+ \Delta t^2 B(\eta u^{n+1,i+1} + (1 - 2\eta)u^n + \eta u^{n-1}),$$

for constants A, B.

A.3.4 Eigenvalue Methods

The eigenvalue methods use the same steps as the standard methods. The difference consists in first determining the stiff and nonstiff operators of the partial differential equation (PDE). In the difference schemes, the stiff operator is then treated implicitly in the first step, and explicitly in the second step. See [102] for further explanations.

A.3.5 Numerical Examples

When starting the program by typing

```
>> startFIDOS;
```

the user is asked to type in the number(s) of the problem as well as of the solver(s). Some problems or solvers can be started together. Then the number of time-steps, or a vector of numbers, as well as the number of space steps, or a vector as well, have to be specified.

When solving a problem for different numbers of time- and space-steps, it is interesting to know the numerical rate of convergence:

$$\rho(\Delta t_1, \Delta t_2) = \frac{\ln(err_{L_1}(\Delta t_1) - err_{L_1}(\Delta t_2))}{\ln(\Delta t_1 - \Delta t_2)}. \tag{A.49}$$

FIDOS calculates this rate automatically, if wished, and gives the whole L_AT_EX-table to be included in a paper with all results and convergence rates as well as label and caption. Subsequently, one program run is given as an example Table A.1 was created automatically.

```
>> startFIDOS;
-----------------------------------------------------------
-----------------------------------------

 FIDOS - solving PDEs with FInite Difference schemes
and Operator-Splitting methods
-----------------------------------------------------------
-----------------------------------------


Problem types:

    [1] wave equation
    [2] viscous Burgers equation
    [3] mixed convection-diffusion Burgers equation
    [4] momentum equation
For more information about the problems, type help, problem and
number of the problem, e.g. >> help problem1

ATTENTION, some methods and/or problems cannot be started
together!

    [1] ADI      for problem  1
    [2] LOD      for problems 1
    [3] IOS      for problems 2,3,4
    [4] EVIOS    for problem  2
    [5] etaIOS   for problem  1
    [6] etaEVIOS for problem  2
For more information about the solvers, type help and name of
the method, e.g. >> help ADI
```

Now the specifications of problem, solver, time-steps, space-steps, tex-table and plots follow:

```
-------------------------------------------------------
 Get problems, solvers and problem-dependent values
-------------------------------------------------------


Problem number(s): 1
Solver number(s): 5

Number(s) of temporal steps on [0,T], e.g. 16
or [4,8]: [4,8,16]
Number(s) of spatial steps on [0,1]^d, e.g. 4
or [4,8]: [4,8,16]

Type 1 to see the tex-tables, otherwise 0: 1
Type 1 to see the plots, otherwise 0: 1
```

The output follows immediately:

```
--------------------
 Start Calculations
--------------------


etaIOS, wave, dx=0.25, dy=0.25, dt=0.3125
Errors at time T: L1-error = 0.048967, max. error = 0.13442

etaIOS, wave, dx=0.125, dy=0.125, dt=0.3125
Errors at time T: L1-error = 0.021581, max. error = 0.054647

etaIOS, wave, dx=0.0625, dy=0.0625, dt=0.3125
Errors at time T: L1-error = 0.014382, max. error = 0.035715

etaIOS, wave, dx=0.25, dy=0.25, dt=0.15625
Errors at time T: L1-error = 0.042097, max. error = 0.11556

etaIOS, wave, dx=0.125, dy=0.125, dt=0.15625
Errors at time T: L1-error = 0.014849, max. error = 0.037601

etaIOS, wave, dx=0.0625, dy=0.0625, dt=0.15625
Errors at time T: L1-error = 0.0077193, max. error = 0.01917

etaIOS, wave, dx=0.25, dy=0.25, dt=0.078125
Errors at time T: L1-error = 0.038395, max. error = 0.1054

etaIOS, wave, dx=0.125, dy=0.125, dt=0.078125
Errors at time T: L1-error = 0.011013, max. error = 0.027888

etaIOS, wave, dx=0.0625, dy=0.0625, dt=0.078125
```

```
Errors at time T: L1-error = 0.0038664, max. error = 0.0096016
```

The L_AT_EX-code for the table and the plots, see Table A.1 and Figure A.3, is given as well.

TABLE A.1: Numerical results for wave equation with coefficients $(D_1, D_2) = (1, 1)$ using etaIOS method, initial condition $U_{n+1,0}(t) = U_n$, and two iterations per time-step.

$\Delta x = \Delta y$	Δt	err_{L_1}	$\text{err}_{\max}$	ρ_{L_1}	$\rho_{\max}$
0.25	0.3125	0.048967	0.13442		
0.125	0.3125	0.021581	0.054647	1.1821	1.2985
0.0625	0.3125	0.014382	0.035715	0.58548	0.61361
0.25	0.15625	0.042097	0.11556		
0.125	0.15625	0.014849	0.037601	1.5034	1.6199
0.0625	0.15625	0.0077193	0.01917	0.94382	0.97195
0.25	0.078125	0.038395	0.1054		
0.125	0.078125	0.011013	0.027888	1.8017	1.9182
0.0625	0.078125	0.0038664	0.0096016	1.5102	1.5383

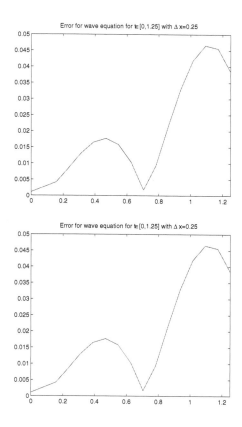

FIGURE A.3: Two of the figures that are automatically created by FIDOS.

Appendix B

Discretization Methods

In this appendix, we discuss the spatial and time discretization methods that we applied in our previous chapters.

B.1 Discretization

In the discretization we deal with the underlying methods to obtain a system of ordinary differential equations.

The underlying idea is to use the spatial discretization methods to transform our partial differential equations into ordinary differential equations. The resulting operator equations can be treated with the decomposition methods and the theory is based on the semigroup analysis.

We concentrate on the Galerkin method to do the spatial discretization.

B.1.1 Galerkin Method

The underlying ideas of transforming the spatial differential terms into operators are given in the Galerkin method.

The method we obtain approximates the *Galerkin solutions* and reaches an operator equation that can be treated as a system of ordinary differential equations.

We have the following method:

- Multiplication of the differential equation for u by the functions $v \in K$ and integration over Ω with subsequent integration by parts yields to the generalized problem.

- The function class K has to be chosen in such a way that a sufficient smooth solution u of the generalized problem is also a solution of the original classical problem in the case where the data are sufficiently smooth (boundary, coefficients, boundary and initial value, etc.).

- Restriction to u and v in the generalized problem to appropriate finite-dimensional subspaces.

The classical derivatives are replaced by generalized (weak) derivatives. The realization of the Galerkin method is done with the basis functions $w_1, \ldots, w_m$, see [200].

In the following, we realize the abstract ideas to the parabolic and hyperbolic differential equations, where the aim is to derive the operator equation. With this abstract generalized problem of coupled systems, we can apply the splitting methods.

B.1.2 Parabolic Differential Equation Applied with the Galerkin Method

In the following we apply the Galerkin method to a parabolic differential equation given as

$$u_t(x,t) - \Delta u(x,t) = f(x,t) , \quad \text{in } \Omega \times [t_0, T] , \tag{B.1}$$
$$u(x,t) = 0 , \quad \text{on } \partial\Omega \times [t_0, T] \text{ (boundary condition)} ,$$
$$u(x,0) = u_0(x) , \quad \text{on } \Omega \text{ (initial condition)}.$$

The generalized problem of Equation (B.1) yields to the following formulation: We seek a function $u \in C([t_0, T]; L^2(\Omega)) \cap C((t_0, T]; H_0^1(\Omega))$, such that

$$\frac{d}{dt} \int_\Omega u(x,t)v(x) \, dx$$
$$+ \int_\Omega \left(\sum_{i=1}^N D_i u(x) \, D_i v(x) - f \, v(x) \right) dx = 0, \tag{B.2}$$
$$u(x,t) = 0 , \quad \text{on } \partial\Omega \times [t_0, T] \text{ (boundary condition)},$$
$$u(x,0) = u_0(x) , \quad \text{on } \Omega \text{ (initial condition)},$$

holds for all $v \in C_0^\infty(\Omega) \cap H_0^1(\Omega)$. Further, we assume $f(x,t) \in L^2([t_0, T]; L^2(\Omega))$ and $u_0(x) \in H_0^1(\Omega)$.

The function v only depends on the space variable x.

In order to construct an approximate solution u_m with the Galerkin method, we make the following attempt:

$$u_m(x,t) = \sum_{k=1}^m c_{k,m}(t) \, w_k(x) , \tag{B.3}$$

where the unknown coefficient $c_{k,n}$ depends on time, and we replace u in the generalized problem (B.2) by u_n and require that (B.2) holds for all $v \in span\{w_1, \ldots, w_m\}$. Then we obtain the Galerkin equations for $j = 1, \ldots, m$:

$$\sum_{k=1}^m c'_{km}(t) \int_\Omega w_k \, w_j \, dx$$
$$+ \sum_{k=1}^m c_{km} \int_\Omega \sum_{i=1}^N D_i w_k \, D_i w_j \, dx = \int_\Omega f \, w_j dx , \tag{B.4}$$

where we have a linear system of first-order ordinary differential equations for the real functions $c_{1m}, \ldots, c_{mm}$. n is the number of the dimension for the test space, and for $m \to \infty$ we obtain the convergence of u_m to u.

Based on this notation we can write in abstract operator equations:

$$Mc_m'(t) + Ac_m(t) = F(t) , \text{ in } [t_0, T], \tag{B.5}$$
$$c_m(0) = \alpha_m, \tag{B.6}$$

where

$$M = \begin{pmatrix} \int_\Omega w_1 \, w_1 \, dx & \ldots & \int_\Omega w_1 \, w_m \, dx \\ \vdots & \ddots & \vdots \\ \int_\Omega w_m \, w_1 \, dx & \ldots & \int_\Omega w_m \, w_m \, dx \end{pmatrix}, \tag{B.7}$$

$$A = \begin{pmatrix} \int_\Omega \sum_{i=1}^N D_i w_1 \, D_i w_1 \, dx & \ldots & \int_\Omega \sum_{i=1}^N D_i w_1 \, D_i w_m \, dx \\ \vdots & \ddots & \vdots \\ \int_\Omega \sum_{i=1}^N D_i w_m \, D_i w_1 \, dx & \ldots & \int_\Omega \sum_{i=1}^N D_i w_m \, D_i w_m \, dx \end{pmatrix}, \tag{B.8}$$

and $F(t) = (\int_\Omega f \, w_1 dx, \ldots, \int_\Omega f \, w_m dx)^T$ are the operators.

The initial conditions $c_m(0)$ correspond to the approximately initial condition of the original problem, so we choose a sequence

$$u_{m,0}(x) = \sum_{k=1}^m \alpha_{km} w_k(x), \tag{B.9}$$

where $u_{m,0}$ converges to u_0 as $m \to \infty$.

For the boundary conditions we have also

$$u_{m,0}(x, t) = 0 , \text{ on } \partial\Omega \times [t_0, T]. \tag{B.10}$$

REMARK B.1 If we replace the linear operators by nonlinear operators in (B.1), we obtain a nonlinear system of ordinary differential equations. This can be treated by linearization and we obtain again our linear ordinary differential equations. ⬜

REMARK B.2 The extensions with convection and reaction terms can also be done with the presented parabolic differential equation. The generalization is done in [81] and [124]. Because of the operator notation as a further simplification of the presented partial differential equations, we concentrate on the diffusion term. ⬜

B.1.3 Hyperbolic Differential Equation Applied with the Galerkin Method

In the following, we apply the Galerkin method to a hyperbolic differential equation given as:

$$u_{tt}(x,t) - \Delta u(x,t) = f(x,t) , \text{ in } \Omega \times [t_0, T] , \qquad (B.11)$$
$$u(x,t) = 0 , \text{ on } \partial\Omega \times [t_0, T] \text{ (boundary condition)},$$
$$u(x,0) = u_0(x) , \text{ on } \Omega \text{ (initial condition)},$$
$$u_t(x,0) = u_1(x) , \text{ on } \Omega,$$

where f, u_0, u_1 are given and we seek u.

The generalized problem of Equation (B.11) yields to the following formulation:

We seek a function $u(x,t) \in C^1([t_0, T]; L^2(\Omega)) \cap C([t_0, T]; H_0^1(\Omega))$, such that

$$\frac{d^2}{dt^2} \int_\Omega u(x,t)v(x) \, dx$$
$$+ \int_\Omega \left(\sum_{i=1}^{N} D_i u(x,t) \, D_i v(x) - f(x,t) \, v(x) \right) dx = 0, \qquad (B.12)$$
$$u(x,t) = 0 , \text{ on } \partial\Omega \times [t_0, T] \text{ (boundary condition)},$$
$$u(x,0) = u_0(x) \text{ on } \Omega \text{ (initial condition)},$$
$$u_t(x,0) = u_1(x) \text{ on } \Omega \text{ (initial condition)},$$

holds for all $v(x) \in C_0^\infty(\Omega) \cap H_0^1(\Omega)$.

Further, we assume $f(x,t) \in L^2([t_0, T]; L^2(\Omega))$, $u_0(x) \in H_0^1(\Omega)$, and $u_1(x) \in L^2(\Omega)$.

The function $v(x)$ depends only on the spatial variable x.

In order to construct an approximate solution u_m with the Galerkin method, we make the following attempt:

$$u_m(x,t) = \sum_{k=1}^{m} c_{k,m}(t) \, w_k(x) , \qquad (B.13)$$

where the unknown coefficient $c_{k,m}$ depends on time, and we replace u in the generalized problem (B.12) by u_m and require that (B.12) holds for all $v \in span\{w_1, \ldots, w_m\}$. Then we obtain the Galerkin equations for $j = 1, \ldots, m$:

$$\sum_{k=1}^{m} c_{km}''(t) \int_\Omega w_k \, w_j \, dx$$
$$+ \sum_{k1}^{m} c_{km} \int_\Omega \sum_{i=1}^{N} D_i w_k \, D_i w_j \, dx = \int_\Omega f \, w_j dx , \qquad (B.14)$$

where we have a linear system of second-order ordinary differential equations for the real functions $c_{1m}, \ldots, c_{mm}$.

Based on this notation we can write in abstract operator equations:

$$Mc_m''(t) + Ac_m(t) = F(t) , \text{ in } [t_0, T], \tag{B.15}$$

$$c_m(0) = \alpha_m, \tag{B.16}$$

$$c_m'(0) = \beta_m, \tag{B.17}$$

where M and A are given as in Equations (B.7) and (B.8).

The initial conditions $c_m(0)$ and $c_m'(0)$ correspond to the approximately initial condition of the original problem, so we choose a sequence

$$u_{m,0}(x) = \sum_{k=1}^{m} \alpha_{km} w_k(x), \tag{B.18}$$

$$u_{m,1}(x) = \sum_{k=1}^{m} \beta_{km} w_k(x), \tag{B.19}$$

where $u_{m,0}$ converges to u_0 and $u_{m,1}$ to u_1 as $m \to \infty$.

For the boundary conditions, we also have

$$u_{m,0}(x,t) = 0 , \text{ on } \partial\Omega \times [t_0, T] . \tag{B.20}$$

REMARK B.3 If we replace the linear operators by nonlinear operators in (B.11), we obtain a nonlinear system of ordinary differential equations. This can be treated by linearization (e.g., Newton method or fixpoint iterations, see [131]), and we obtain again our linear ordinary differential equations.
□

B.1.4 Operator Equation

For more abstract treatment of our splitting methods in Chapter 3, we deal in the following with the operator equations:

$$A_1 \frac{d^2}{dt^2} C_1(u(t), t) + A_2 \frac{d}{dt} C_2(u(t), t) + B(u(t), t) = 0 , \; \forall t \in [t_0, T], \tag{B.21}$$

where A_1, A_2, B, C_1, C_2 are positive definite and symmetric operators, so at least also boundable operators for all $t \in [t_0, T]$.

We have also reset the notation c_n to u for the abstract treatment. At least the results have to be retransformed with the test-space, see Equation (B.13).

For the operator equations we can in the following chapters treat our underlying splitting methods with the semigroup theory.

B.1.5   Semigroup Theory

With the notion of a semigroup we can describe time-dependent processes in nature in terms of the functional analysis. We describe an introduction that is needed in the further sections. An overview to the semigroup theory can be found in [12], [13], [65], [191], [198], [200], and [201].

The key relations are

$$S(t + s) = S(t)S(s) \ , \ \forall t, s \in \mathbb{R}^+ \ , \tag{B.22}$$
$$S(0) = I \ , \tag{B.23}$$

and we have the following definition of a generator of a semigroup.

DEFINITION B.1 A semigroup $\{S(t)\}$ on a Banach space X consists of a family of operators $S(t) : X \to X$ for all $t \in \mathbb{R}^+$ with B.22 and B.23. The generator $B : D(B) \subset X \to X$ of the semigroup $\{S(t)\}$ is defined by

$$Bw = \lim_{t \to 0^+} \frac{S(t)w - w}{t} \ , \tag{B.24}$$

where w belongs to $D(B)$, if the limit of B.24 exists.

A one-parameter group $\{S(t)\}$ on the Banach space X consists of a family of operators $S(t) : X \to X$ for all $t \in \mathbb{R}$, with (B.22) for all $t, s \in \mathbb{R}$ and (B.23) as the initial condition.

REMARK B.4 A family $\{S(t)\}$ of linear continuous operators from X to itself, which satisfies the conditions with (B.22) and (B.23), is called a continuous semigroup of linear operators or simply a C_0 semigroup. ▯

B.1.6   Classification of Semigroups

Let $\mathcal{S} = \{S(t)\}$ be a semigroup of the Banach space X.
We have the following classification of the semigroups:

1. $\mathcal{S}$ is called strongly continuous, if $t \mapsto S(t)w$ is continuous on $\mathbb{R}^+$ for all $w \in X$ that is,

$$\lim_{t \to s} S(t)w = S(s)w \ , \ \forall s \in \mathbb{R}^+. \tag{B.25}$$

2. $\mathcal{S}$ is called uniformly continuous, if all operators $S(t) : X \to X$ are linear and continuous, and $t \mapsto S(t)$ is continuous on $\mathbb{R}^+$ with respect to the operator norm that is,

$$\lim_{t \to s} ||S(t) - S(s)|| = 0 \ , \ \forall s \in \mathbb{R}^+. \tag{B.26}$$

3. S is called nonexpansive, if all operators $S(t) : X \to X$ are nonexpansive and

$$\lim_{t \to 0^+} S(t)w = w , \ \forall w \in X. \tag{B.27}$$

4. S is called a linear semigroup, if all operators $S(t) : X \to X$ are linear and continuous.

5. S is called an analytical semigroup, if we have an open sector Σ and a family of linear continuous operators $S(t) : X \to X$ for all $t \in \Sigma$ with $S(0) = I$ and the following properties:
   (a) $t \mapsto S(t)$ is an analytical map from Σ into $L(X,Y)$,
   (b) $S(t + s) = S(t)S(s)$ for all $t, s \in \Sigma$,
   (c) $\lim_{t \to 0^+} S(t)w = w$ in Σ for all $w \in X$,
   where an open sector is defined as

$$\Sigma = \{z \in \mathbb{C} : -\alpha < \arg(z) < \alpha , \ z \neq 0\}, \tag{B.28}$$

   see also Figure B.1.

6. S is called a bounded analytical semigroup if we have an analytical semigroup and additionally the property for each $\beta \in \,]0, \alpha[$ is satisfied:

$$\sup_{t \in \Sigma_\beta} ||S(t)|| < \infty, \tag{B.29}$$

where $\Sigma_\beta = \{z \in \mathbb{C} : -\beta < \arg z < \beta , \ z \neq 0\}$.

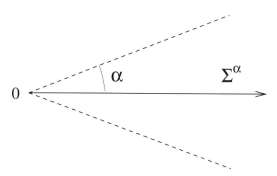

FIGURE B.1: Sector for the analytical semigroup.

The following examples are given for the semigroups.

Example B.1

(*i*) If $B : D(B) \subset X \to X$ is a linear self-adjoint operator on the Hilbert space X with $(Bu, u) \le 0$ on $D(B)$, then B is the generator of a linear nonexansive semigroup. Such semigroups describe, in particular, heat conduction and diffusion processes. In terms of the general functional calculus for self-adjoint operators, this semigroup is given by $\{\exp(tB)\}$.

(*ii*) If $H : D(H) \subset X \to X$ is a linear self-adjoint operator on the complex Hilbert space X, then $-iH$ generates an one-parameter unitary group. Such groups describe the dynamics of quantum systems. The operator H corresponds with the energy of the quantum system and is called Hamiltonian of the system. In terms of the general functional calculus for self-adjoint operators on the complex space, this semigroup is given by $\{\exp(-itH)\}$.

(*iii*) If $C : D(C) \subset X \to X$ is a skew-adjoint operator on the real Hilbert space X, then C is the generator of a one-parameter unitary group. Such semigroups describe, for example, the dynamics of wave processes. ▯

In our monograph we will treat examples (*i*) and (*iii*) that is, the self-adjoint and the skew-adjoint operator on the real Hilbert space X.

Further, for our realistic application in heat equations, we have to assume unbounded operators because of the irreversibility of the processes, see also [200].

B.1.7 Abstract Linear Parabolic Equations

For the discussion of the parabolic equations we consider the equations in a notation of an abstract initial value problem given as

$$u'(t) = Bu(t) + f(t) , \ t_0 < t < T, \tag{B.30}$$
$$u(0) = w,$$

and the solution of (B.30):

$$u(t) = S(t - t_0) \, w + \int_{t_0}^{t} S(t - s) \, f(s) \, ds. \tag{B.31}$$

where the integration term is a convolution integral, see [200], and can be solved numerically with Runge-Kutta methods, see [116] and [117].

We also have the following assumptions.

Assumption B.1 (*H*1) *Let $\{S(t)\}$ be a strongly continuous linear semigroup on the Banach space X over $\mathbb{R}$ or $\mathbb{C}$ with the generator B that is, $\{S(t)\}$ is a semigroup of linear continuous operators $S(t) : X \to X$ for all $t \ge 0$, and $t \mapsto S(t)w$ is continuous on $\mathbb{R}^+$ for all $w \in X$.*

(*H*2) *The function $f : [t_0, T[\to X$ is continuous.*

THEOREM B.1
Assume (H1), (H2). Then there holds:

(a) *There exists at most one classical solution of* (B.30), *and each classical solution is also a mild solution (weak solution).*

(b) *If $f \in C^1$ and $w \in D(B)$, then the mild solution* (B.31) *is also a classical solution of* (B.30).

(c) *If the operator $B : X \to X$ is linear and continuous, then, for each $w \in S$ and each continuous f, the mild solution* (B.31) *is also a classical solution of* (B.30).

Therefore we have the importance of the semigroups for the solution of the initial value problem (B.30).

COROLLARY B.1

(i) *There exist constants $C \geq 1$ and $a \geq 0$, such that*

$$||S(t)|| \leq C \, \exp(at) \, , \, \forall \, t \geq 0 \, . \qquad (B.32)$$

(ii) *The generator $B : D(B) \subset X \to X$ of the semigroup $\{S(t)\}$ is a linear graph closed operator and $D(B)$ is dense in X.*

(iii) *The semigroup is uniquely determined by its generator.*

With this notation of the semigroup for the parabolic equation, we can abstractly treat the splitting methods, see Chapter 3.

In the next subsection we discuss the abstract hyperbolic equation.

B.1.8   Abstract Linear Hyperbolic Equations

For the discussion of the hyperbolic equations, we consider the equations in a notation of an abstract initial value problem given as

$$u''(t) + Au(t) = f(u(t)) \, , \, 0 < t < \infty \, , \qquad (B.33)$$
$$u(0) = u_0 \, , \, u'(0) = u_1 \, .$$

We also have the following assumptions.

Assumption B.2 *(H1) The linear operator $A : D(A) \subset X \to X$ is self-adjoint and strongly monotone on the Hilbert space over $\mathbb{R}$ or $\mathbb{C}$. Let X_E be the energetic space of A with the norm $|| \cdot ||_E$ that is, X_E is* *the completion of $D(A)$ with respect to the energetic scalar product $(u, v)_E = (Au, v)$. So in other words, A is a symmetric and positive definite operator, see also [65].*

(H2) *The operator $f : X_E \rightarrow X$ is locally Lipschitz continuous that is, for each $R > 0$ there is a constant L such that $\|f(u) - f(v)\| \leq L\|u - v\|_E$, for all $u, v \in X_E$ with $\|u\|_E, \|v\|_E \leq R$.*

We set $v = u'$. We rewrite the equation (B.33) into a first-order system and achieve

$$\begin{pmatrix} u' \\ v' \end{pmatrix} = \begin{pmatrix} 0 & I \\ -A & 0 \end{pmatrix} \begin{pmatrix} u \\ v \end{pmatrix} + \begin{pmatrix} 0 \\ f \end{pmatrix}. \tag{B.34}$$

We can set $z = (u, v)$ and rewrite (B.34) in the form

$$z''(t) = Cz(t) + F(z(t)), \ 0 < t < \infty, \tag{B.35}$$
$$z(0) = z_0.$$

Let $Z = X_E \times X$ and $D(C) = D(A) \times X_E$.

If we use the assumption $(H1)$, then the operator C is skew-adjoint and generates a one-parameter unitary group $\{S(t)\}$.

The applications to this semigroup are discussed in Chapter 4.5, and we discuss the iterative splitting method with respect to consistency and stability analysis in the semigroup notation.

Furthermore, for many applications nonlinear semigroups are important. So we describe in the next subsection the notations and important results for the abstract nonlinear semigroup theorem that we need in the next chapters.

B.1.9 Nonlinear Equations

In this section we discuss the abstract semigroup theory for linear operators by introducing certain nonlinear semigroups, generated by convex functions, see [65] and [166]. They can be applied for various nonlinear second-order parabolic partial differential equations.

In the following we apply the following nonlinear semigroups to our nonlinear differential equations.

We have a Hilbert space H and take $I : H \rightarrow (-\infty, +\infty]$ to be convex, proper, lower semicontinuous.

For simplicity we assume as well that ∂I is densely defined that is, $\overline{D(\partial I)} = H$.

Further we propose to study the nonlinear differential equation given as

$$u'(t) + A(u(t)) \ni 0, \ 0 \leq t < \infty, \tag{B.36}$$
$$u(0) = u_0,$$

where $u \in H$ is given and $A = \partial I$ is a nonlinear, discontinuous operator and is also multivalued.

For the convex analysis, we also assume that (B.36) has a unique solution for each initial point u. We write

$$u(t) = S(t)u = 0 \ , \ 0 < t < \infty, \tag{B.37}$$

and regard $S(t)$ as a mapping from H into H for each time point $t \geq 0$.

We note that the mapping $u \mapsto S(t)u$ is, in general, nonlinear.

As we defined the linear semigroup we also have further conditions, see [65].

$$S(0)u = u \ , \ \text{for } u \in H, \tag{B.38}$$
$$S(t+s)u = S(t)S(s)u \ , \ t,s \geq 0, \ u \in H, \tag{B.39}$$
the mapping $t \mapsto S(t)u$ is continuous from $[0,\infty)$ into H.

Then we define the nonlinear semigroup as follows.

DEFINITION B.2

(i) A family $\{S(t)\}$ of nonlinear operator mappings H into H is called a nonlinear semigroup, if the conditions (B.38)–(B.40) are satisfied.

(ii) We say $\{S(t)\}$ is a contraction semigroup, if in addition there holds

$$||S(t)u - S(t)\hat{u}|| \leq ||u - \hat{u}|| \ , \ t \geq 0, u, \hat{u} \in H. \tag{B.40}$$

We could show that the operator $A = \partial I$ generates a nonlinear semigroup of contractions on H.

So we could solve the ordinary differential equation:

$$u'(t) \in \partial I(u(t)) \ , \ t \geq 0, \tag{B.41}$$
$$u(0) = u_0, \tag{B.42}$$

which is well posed for a given initial point $u \in H$. This is done by a kind of infinite dimensional gradient flow, see [65].

For the theory, the idea in our monograph is to regularize or smooth the operator $A = \partial I$ with linearization (e.g., Taylor expansions or fixpoint iterations with previous iterates), or finding an A_λ with a resolvent as a regularization.

REMARK B.5 For the regularization, the nonlinear resolvent J_λ can be defined as:

DEFINITION B.3

(1) For each $\lambda > 0$ we define the nonlinear resolvent $J_\lambda : H \to D(\partial I)$ by setting

$$J_\lambda[w] := u ,$$

where u is a unique solution of

$$w \in u + \lambda \partial I[u] .$$

(2) For each $\lambda > 0$ we define the Yoshida approximation $A_\lambda : H \to H$ by

$$A_\lambda[w] := \frac{w - J_\lambda[w]}{\lambda} , \quad w \in H . \qquad (B.43)$$

So therefore A_λ is a sort of regularization or smoothing of the operator $A = \partial I$. □

For the semigroup theory we can define the resolvent with the infinitesimal generator A, see [198].

THEOREM B.2
If $n > 0$, then the operator $(nI - A)$ admits an inverse
$$R(n, A) = (nI - A)^{-1} \in L(X, X), \text{ and}$$

$$R(n, A)x = \int_0^\infty \exp(-ns) \, T_s x \, ds \text{ for } x \in X , \qquad (B.44)$$

where T_s is the linear operator of the semigroups.
   Thus, positive real numbers belong to the resolvent set $\rho(A)$ of A and we have a sort of eigenvalue for the inverse operator.

These can be used to study the nonlinear problems with the help of the eigenvalue problems.
   In the next sections we discuss our underlying time- and space-discretization methods for our coupled systems of parabolic and hyperbolic differential equations.

B.2   Time-Discretization Methods

B.2.1   Introduction

In this section we concentrate on the time-discretization methods for ordinary differential equations. We assume that we have semi-discretized our partial differential equations to ordinary differential equations so that we could start on the level of ordinary differential equations.

The spatial discretization of the equations is done with respect to the type of the differential equations and is described in Section B.3.

We focus on the discretization of the convection-diffusion-reaction and the diffusion reaction equation and introduce improved discretization methods for extreme parameters that lead to stiff equations. It is interesting to discuss the standard methods and to enrich them with analytical or semi-analytical methods.

The idea behind the discretization methods in time and space is to decouple the spatiotemporal discretization.

Often decoupling methods are applied after discretizing time and space variables. Here the balance between the time and space discretization methods is important. So the spatiotemporal schemes can be balanced with higher-order discretization schemes and implicit-explicit time discretization methods, see [2].

The decoupling in time and space has the advantage of more efficiency and acceleration.

In the next sections we discuss the discretization methods for the time variable and for the space variables.

B.2.2 Discretization Methods for Ordinary Differential Equations

Based on the decoupling idea in time and space, the semi-discretization of the spatial terms leads to an ordinary differential equation for the time variable.

Thus, the study of the behavior of an ordinary differential equation (ODE) is important for the application of the method.

We propose the semi-discretization idea of decoupling in time and space and treating the time variable with discretization methods of ODEs.

We deal with a system of ordinary differential equations:

$$y_1' = f_1(y_1, \ldots, y_n), \tag{B.45}$$

$$\vdots$$

$$y_n' = f_n(y_1, \ldots, y_n),$$

where the initial conditions are $(y_1(0), \ldots, y_n(0))^t = (y_{1,0}, \ldots, y_{n,0})^t$.
The functions $f_1, \ldots, f_n$ are derived from the spatial discretization.

Based on this equation we choose our discretization methods for the time variable.

First we characterize our system as a stiff or nonstiff problem.

We define a system of ODEs as a stiff system.

DEFINITION B.4 An initial value problem is stiff, if we have the following condition:

$$(t_e - t_0)||\frac{\partial f(t, y)}{\partial y}|| >> 1, \text{ or} \tag{B.46}$$

$$L(t_e - t_0) >> 1,$$

where L is the classical Lipschitz constant and t_0 and t_e are the initial and end time, respectively.

This classification is important for the further design of discretizations and solver methods.

For testing the discretization and solver methods, we introduce the test equation:

$$y' = \lambda y, \tag{B.47}$$

$$y(0) = y_0,$$

where $Re(\lambda) \leq 0$. This is called the Dahlquist test problem.

The idea behind, is to test the discretization methods with this simple test example and to represent the stiff problem.

We could then discuss the stability of a method. Therefore we define the A-stability and L-stability.

DEFINITION B.5 A one-level method is A-stable, if we have

$$|R_0(z)| \leq 1, \ \forall z \in \mathbb{C} \text{ with } Re(z) \leq 0, \tag{B.48}$$

where $R_0(z)$ is the stability function of the temporal discretization. In this case the stability function is called A-acceptable.

DEFINITION B.6 A one-level method is L-stable, if it is A-stable and if

$$\lim_{Re(z) \to -\infty} R_0(z) = 0. \tag{B.49}$$

If it satisfies only the weak condition

$$\lim_{Re(z) \to -\infty} |R_0(z)| < 1, \tag{B.50}$$

we call it strong A-stable.

B.2.2.1 Nonstiff System of ODEs

For the discretization of nonstiff systems of ODEs we introduce the following discretization methods:

1) Explicit one-level methods: We discuss the explicit Runge-Kutta methods.

2) Explicit multilevel methods: We discuss the predictor-corrector methods.

B.2.2.2   Explicit One-Level Methods

One-level methods are defined as methods that are solved in one level, see [116].

We define the explicit Runge-Kutta method as representative of the one-level methods.

DEFINITION B.7 Let $s \in \mathbb{N}$. The one-level method of the structure

$$u_{m+1} = u_m + h \sum_{i=1}^{s} b_i f(t_m + c_i h, u_{m+1}^{(i)}), \tag{B.51}$$

$$u_{m+1}^{(i)} = u_m + h \sum_{j=1}^{i-1} a_{ij} f(t_m + c_j h, u_{m+1}^{(j)}), \tag{B.52}$$

$$i = 1, \ldots, s,$$

is an s-level implicit Runge-Kutta method. We choose the parameters a_{ij}, b_i, and c_i in such a manner that we get a stable method.

B.2.2.3   Explicit Multilevel Methods

We define the following multilevel methods.

DEFINITION B.8 An general multilevel method is of the form

$$\sum_{l=0}^{k} \alpha_l y_{m+l} = h \sum_{l=0}^{k-1} \beta_l f(t_{m+l}, u_{m+l}). \tag{B.53}$$

B.2.2.4   Stiff System of ODEs

For the discretization of stiff systems of ODEs we introduce the following discretization methods:

1) Implicit one-level methods: We discuss the implicit Runge-Kutta methods.

2) Implicit multilevel methods: We discuss the BDF methods (backward differential methods).

B.2.2.5 Implicit One-Level Methods

We define the implicit Runge-Kutta method as a representative of the one-level methods.

> **DEFINITION B.9** Let $s \in \mathbb{N}$. The one-level method of the structure

$$u_{m+1} = u_m + h \sum_{i=1}^{s} b_i f(t_m + c_i h, u_{m+1}^{(i)}), \qquad \text{(B.54)}$$

$$u_{m+1}^{(i)} = u_m + h \sum_{j=1}^{s} a_{ij} f(t_m + c_j h, u_{m+1}^{(j)}), \qquad \text{(B.55)}$$

$$i = 1, \ldots, s,$$

is an s-level implicit Runge-Kutta method. We choose the parameters $a_{ij}, b_i,$ and c_i in such a manner, that we get a stable method.

B.2.2.6 Implicit Multilevel Methods

We define the following multilevel methods.

> **DEFINITION B.10** A general multilevel method is of the form

$$\sum_{l=0}^{k} \alpha_l y_{m+l} = h \sum_{l=0}^{k} \beta_l f(t_{m+l}, u_{m+l}). \qquad \text{(B.56)}$$

We discuss in the following sections the BDF methods as representative of the implicit multilevel method.

B.2.3 Characterization of a Stiff Problem

For the discretization methods, a further qualification and notation of the character of the equations are necessary to design stable methods. For example, in chemical reactions, very fast-reacting processes lead to an equilibrium, which is known as a stiff problem.

B.2.3.1 Stability Function of a RK Method

> **DEFINITION B.11** The stability function of an implicit Runge-Kutta method is defined as

$$R_0(z) = \frac{det(I - zA + z \mathbb{1} b^t)}{det(I - zA)}, \qquad \text{(B.57)}$$

where b, A, and z are given as in the usual Runge-Kutta denotation in the matrix-tableau form.

B.2.3.2 Error Estimate: Local and Global Error

The local error of the implicit Runge-Kutta methods is defined as

$$\delta_h(x) = y_1 - y(x+h). \tag{B.58}$$

REMARK B.6

1. If $\alpha_0(A^{-1}) > 0$, then there holds

$$||\delta_h(x)|| \le C \, h^{q+1} \max_{\xi \in [x, x+h]} ||y^{(q+1)}(\xi)||, \tag{B.59}$$

$$\text{for } h \, \mathbf{v} \le \alpha \le \alpha_0(A^{-1}),$$

where α_0 is the stiffness function of the test problem, $\mathbf{v}$ is a constant, h is the time-step, and A is the matrix of the RK method.

2. If $\alpha_D(A^{-1}) = 0$ for some positive diagonal matrix D and $\mathbf{v} < 0$, then

$$||\delta_h(x)|| \le C \, (h + \frac{1}{|\mathbf{v}|}) h^q \max_{\xi \in [x, x+h]} ||y^{(q+1)}(\xi)||, \tag{B.60}$$

$$\text{for } h \, \mathbf{v} \le \alpha_D \le \alpha_0(A^{-1}).$$

In both cases the constant C depends only on the coefficients of the Runge-Kutta matrix and on a constant α like in Remark B.6.1.

☐

B.2.3.3 Multilevel Methods

The multilevel methods are used in applications for more accuracy. In the initialization phase the previous values have to be computed. The implicit methods are used for solving stiff problems (e.g., BDF methods).

REMARK B.7 A general multilevel method is given in the following form:

$$\sum_{l=0}^{k} \alpha_l y_{m+l} = h \sum_{l=0}^{k} \beta_l f(t_{m+l}, y_{m+l}), \tag{B.61}$$

where $\alpha_l, \beta_l \in \mathbb{R}^+$ are fixed coefficients to the method.

☐

We can obtain a method of order m if

$$\sum_{i=1}^{k} \alpha_i = 1, \tag{B.62}$$

$$\sum_{i=1}^{k} i\,\alpha_i = \sum_{i=0}^{k} \beta_i, \tag{B.63}$$

$$\sum_{i=1}^{k} i^j\, \alpha_i = j \sum_{i=1}^{k} i^{j-1} \beta_i \ , \ j = 2, 3, \ldots, m. \tag{B.64}$$

PROOF

See the proof outline in [33]. ▯

B.2.3.4 Characteristic Polynomials

The k-step method can be written as

$$\rho(E_h)u_m = h\sigma(E_h)f(t_m, u_m), \tag{B.65}$$

where h is the grid width, $\rho(E_h)$ is the polynomial of the domain, and $\sigma(E_h)$ is the polynomial of the multilevel method.

B.2.3.5 Stability of Multilevel Methods

We apply the test equation

$$y' = \lambda y \ , \ y(0) = y_0, \tag{B.66}$$

to the general explicit or implicit multilevel method and get the following homogeneous difference equation:

$$\sum_{l=0}^{k} \alpha_l y_{m+l} = z \sum_{l=0}^{k} \beta_l f(t_{m+l}, y_{m+l}) \tag{B.67}$$

$$\text{with } z = \lambda h.$$

If we use $u_m = \xi_m$, we get the characteristic polynomial $\rho(\xi_m) - z\sigma(\xi_m) = 0$.

B.2.3.6 Stability Region and A-Stability

We define the stability region of a multilevel method

DEFINITION B.12 The stability region is given as

$$S = \{z \in \mathbb{C} : \text{for all roots } \xi_l \text{ of } \rho(\xi_l) - z\sigma(\xi_l) = 0 \ \text{ we have } |\xi_l| \le 1 \tag{B.68}$$
$$\text{for } \xi_l \text{ multi-roots we have } |\xi_l| < 0\}\,.$$

We have A-stability, if the linear multilevel method satisfies the following definition.

DEFINITION B.13 The A-stability of a linear multilevel method is given, if

$$\mathbb{C}^- \subset S, \tag{B.69}$$

and $A(\alpha)$-stability is given for an $\alpha > 0$, if

$$\{z \in \mathbb{C} \text{ with } |arg(z) - \pi| \leq |\alpha|\} \subset S. \tag{B.70}$$

B.2.3.7 Stability for Explicit Multilevel Methods

A-stability is not satisfied for the explicit methods. So we could not apply the methods for stiff problems.

REMARK B.8 The characteristic equation for the explicit method is given as

$$\alpha_k \xi^k = (\alpha_{k-1} - z\beta_{k-1})\xi^{k-1} + \ldots + (\alpha_0 - z\beta_0) = 0. \tag{B.71}$$

It exists one $\beta_l \neq 0$, where the coefficient of ξ^l goes to infinity with $|z| \to \infty$, and also exists one root. Therefore, we cannot satisfy Definition B.13, and we do not obtain A-stability. ◻

B.2.3.8 BDF Methods

The motivation is to construct implicit methods that satisfy the A-stability. The BDF methods are a good class and they are constructed as

$$\alpha_k y_{m+k} + \alpha_{k-1} y_{m+k-1} + \ldots + \alpha_0 y_m = h f_{m+k}, \tag{B.72}$$

where $\sum_{j=0}^k \alpha_j = 0$ and the coefficients are given as

$$\sum_{j=0}^k \alpha_j W_{m+j} = \sum_{j=0}^k (-1)^j \binom{-s+1}{j} \nabla^j y_{n+1}, \tag{B.73}$$

see also [116].

Because of the construction of the methods, we obtain for the stiff problems the L-stability with $R_0(\infty) = 0$.

Therefore, all roots have the limit 0, and for all k there holds,

$$\rho(\xi) = \beta_k \xi_k \text{ and } \beta_k \neq 0. \tag{B.74}$$

Therefore, the method is constructed with $\beta_0 = \ldots = \beta_{k-1} = 0$.

Typical BDF methods are given as

$$k = 1 : \quad hf_{m+1} = u_{m+1} - u_m, \tag{B.75}$$

$$k = 2 : \quad hf_{m+2} = 0.5(3u_{m+2} - 4u_{m+1} + u_m), \tag{B.76}$$

$$k = 3 : \quad hf_{m+3} = \frac{1}{6}(11u_{m+3} - 18u_{m+2} + 9u_{m+1} - 2u_m). \tag{B.77}$$

B.2.3.9 IMEX Methods

In the field of stiff problems in the 1980s the $IMEX$ implicit-explicit methods were established in the direction of stiff and nonstiff problems, [171] and [172]. For many applications the natural splitting is done into two parts, one of these is nonstiff and suited for explicit treatment, the other stiff one can be handled implicitly. The mixture of implicit and explicit methods has the advantage to solve one method with different operators and take into account the previous information w_{n-j}.

IMEX θ-Method

We deal with the semi-discretized system given as

$$w' = F(t, w(t)) = F_0(t, w(t)) + F_1(t, w(t)), \tag{B.78}$$

where F_0 is the nonstiff term discretized with the explicit method and F_1 is the stiff term discretized with the implicit method.

We can use the following simple θ-stepping method:

$$w_{n+1} = w_n + \theta F_0(t_n, w_n) + (1 - \theta)F_1(t_n, w_n) + \theta F_1(t_{n+1}, w_{n+1}), \tag{B.79}$$

with $\theta \geq 1/2$. We combine an explicit Euler method with an A-stable implicit θ-method.

SBDF Method (Semi-Explicit Backward Differential Formula)

The SBDF methods are additional methods for stiff and nonstiff parts of an evolution equation. So the mixtures of explicit and implicit methods are based on the BDF methods and are discussed in [6] and [7].

The SBDF method is constructed as

$$\sum_{j=0}^{k} \alpha_j W_{m+j} = h \left(\sum_{j=0}^{k} \beta_j F_0(W_{m+j}) + \sum_{j=0}^{k} \gamma_j F_1(W_{m+j}) \right), \tag{B.80}$$

where F_0 is the nonstiff operator and F_1 is the stiff operator. We have further $\sum_{j=0}^{k} \alpha_j = 0$, $\sum_{j=0}^{k} \beta_j = 1$, where $\beta_k = 0$ and $\sum_{j=0}^{k} \gamma_j = 1$ and the construction of the α coefficients are given as

$$\sum_{j=0}^{k} \alpha_j W_{m+j} = \sum_{j=0}^{k} (-1)^j \binom{-s+1}{j} \nabla^j y_{n+1},$$

see also [116].

An example of such methods is for a SBDF-2 (second-order semi-explicit backward differential formula) method the CNAB (Cank-Nicolson, Adam-Bashforth) method, a combination of the implicit-explicit Crank-Nicolson and the explicit Adam-Bashforth method.

For example

$$\frac{W^{n+1} - W^n}{\Delta t} = \frac{3}{2}F_0(W^n) - \frac{1}{2}F_0(W^{n-1}) + \frac{1}{2}F_1(W^{n+1}) + \frac{1}{2}F_1(W^n) \tag{B.81}$$

where F_0 is the nonstiff term discretized with the explicit method and F_1 is the stiff term discretized with the implicit method.

Another explicit-implicit method is the additive Runge-Kutta method.

Additive Runge-Kutta Methods

For more than one operator of different behavior, the class of additive Runge-Kutta methods is established.

We deal with the following semi-discretized system given as

$$y' = f_1(t, y(t)) + \ldots + f_N(t, y(t)), \tag{B.82}$$

$$y(0) = y_0, \tag{B.83}$$

where $f_v : \mathbb{R}_0^+ \times \mathbb{R}^D \to \mathbb{R}^D$, $v = 1, \ldots, N$, are vectorial functions with the components $f_{v,i}$, $i = 1, \ldots, D$.

DEFINITION B.14 An additive Runge-Kutta (ARK) method of s stages and N levels is a one-step numerical method that, for a known approximation y_n to $y(t^n)$, obtains the approximations y^{n+1}, with $t^{n+1} = t^n + h$ with h being the step size, according to the process

$$\begin{cases} Y_{n,i} = y_n + h \sum_{v=1}^{N} \sum_{j=1}^{s} a_{ij}^{[v]} f^{[v]}(t_n + c_j h, Y_{n,j}) \\ y_{n+1} = y_n + h \sum_{v=1}^{N} \sum_{i=1}^{s} b_i^{[v]} f^{[v]}(t_n + c_i h, Y_{n,i}). \end{cases} \tag{B.84}$$

The coefficients of the method may be organized in the Butcher tableau

$$\begin{array}{c|c|c|c|c} c & A^{[1]} & A^{[2]} & \ldots & A^{[N]} \\ \hline & b^{[1]T} & b^{[2]T} & \ldots & b^{[N]T} \end{array}, \tag{B.85}$$

where $c = [c_1, \ldots, c_s]^T$ and, for $v = 1, \ldots, N$, $b^{[v]} = [b_1^{[v]}, \ldots, b_s^{[v]}]^T$ and $A^{[v]} = \left(a_{ij}^{[v]}\right)_{i,j}^s$.

An important subclass of the ARK methods is the class of fractional step Runge-Kutta methods defined as follows.

DEFINITION B.15 A fractional step Runge-Kutta method (FSRK) is an ARK method, see Equation (B.84), which satisfies

1. $a_{ii[v]} \leq 0$, for $i = 1, \ldots, s$, and $v = 1, \ldots, N$, and $a_{ij}^{[v]} = 0$, for all $j > i$.

2. $|b_j^{[v]}| + \sum_{i=1}^{s} |a_{ij[v]}| = 0 \rightarrow |b_j^{[\mu]}| + \sum_{i=1}^{s} |a_{ij[\mu]}| \neq 0$ for $v, \mu = 1, \ldots, N$ such that $\mu \neq v$, and $i, j = 1, \ldots, s$.

3. $a_{ii[\mu]} a_{ii[v]} = 0$ for $v, \mu = 1, \ldots, N$ such that $\mu \neq v$, and $i, j = 1, \ldots, s$.

So the method can be written in one tableau.

The most interesting applications are the ones where every operator is stiff. In this case we can use the methods for stiff solvers.

REMARK B.9 The ARK methods can be applied for ODEs that contain stiff and nonstiff parts. In detecting the most adequate method, an automatization has been developed, see [170]. □

In the next section we discuss the time discretization methods for linear and nonlinear problems.

B.2.4 Linear Nonlinear Problem

For the discretization methods a further improvement can be achieved by designing a time-discretization method for different characters of the underlying equations. Often the mixture of linear and nonlinear problems appears in chemical reactions or biological processes.

Based on the two effects, we could assume a very fast process treated with the nonlinear part and a slow process treated with the linear part that is the stiff part of the equation.

Thus, these two processes can be treated with different discretization methods of higher order, see [6].

The nonlinear part is treated with an explicit method, while the linear part is treated with an implicit method.

$$y' = A(u) + Bu, \qquad (B.86)$$
$$y(0) = y_0,$$

where $A(u)$ is the nonlinear function and B is the linear function.

For such equations the mixtures between implicit and explicit methods are interesting. The explicit part is much simpler to compute and therefore the stiff linear operator B is handled implicitly in order to obtain larger time-steps.

Stable and higher-order methods for such mixed problems lead to large time-steps and fast computations.

In the next subsection we discuss the nonlinear stability.

B.2.4.1 Nonlinear Stability

A further generalization of A-stability for nonlinear equations is given as

$$y'(x) = f(x, y(x)), \tag{B.87}$$
$$y(x_0) = y_0.$$

So the stability is denoted as

$$\langle f(u) - f(v), u - v \rangle \leq 0, \tag{B.88}$$
$$y(0) = y_0,$$

where $\langle \cdot \rangle$ denotes a semi-inner product, with corresponding semi-norm $\|u\| = \langle u, u \rangle^{1/2}$.

We define the nonlinear stability as follows.

DEFINITION B.16 A time-discretization method is BN-stable, if for any initial value problem

$$y'(x) = f(x, y(x)), \ y(x_0) = y_0, \tag{B.89}$$

satisfying the condition $< f(x, u), u > \leq 0$, the sequence of computed solutions satisfies $\|y_n\| \leq \|y_{n-1}\|$.

REMARK B.10 The time-discretization methods can be applied for the nonlinear parts, because the linearization is done. One could further treat the equation parts as stiff and nonstiff operators and apply the ARK-, FSRK-methods, see applications in [132] and [170]. The stability and consistency proofs for the nonlinear parts can be found in [34]. □

B.3 Space-Discretization Methods

B.3.1 Introduction

In this chapter we will focus on the spatial discretization methods for hyperbolic and parabolic partial differential equations.

We concentrate on the finite element and finite volume methods and specialize our discretizations to large-scale equations. With respect to large equation parameters we discuss the discretization schemes, for example for the convection-diffusion reaction equations. Therefore, the standard methods (e.g., finite volume methods), can be enriched with analytical or semianalytical methods and more accurate solutions can be obtained, see [15], [85], and [156].

A further idea behind the spatial discretization methods are the design of methods for discontinuous solutions, so we also discuss the design of such methods, for example, discontinuous Galerkin methods, see [5]. Also in respect of the different spatial scales in the domains and to save computational time, adaptivity is important, see [10], [143], and [192].

In the next sections we discuss the spatial discretization methods and concentrate on first- and second-order spatial derivates, for example, applied on convection-diffusion equations.

The major part of the spatial discretization can be done in the context of elliptic equations, see [46].

We follow the standard ideas of the decoupled discretization of time and space. Therefore, we discuss the general ideas of the finite difference, finite element, and finite volume methods.

For the finite difference methods, the discussion of the approximation of the difference scheme is done, see [45]. The finite element and finite volume methods are introduced by the application of the weak formulations, see [46].

So at least the idea, for the finite discretization methods are the digitalization of domains into finite pieces of local domains, which represents the approximated solutions of the exact solution on the continuous domain.

For a first overview we will start with the elliptic equations and apply the spatial discretization methods.

B.3.2 Model Problem: Elliptic Equation

For a brief introduction, we focus on the elliptic equation of second order, see [65], in the following context:

$$Lu := -u'' + b(x)\,u' + c(x)\,u = f(x) \text{ in } x \in \Omega = (0,1), \quad \text{(B.90)}$$
$$\text{with } u(0) = u_0, \ u(1) = u_1, \quad \text{(B.91)}$$

where the coefficients $b(x)$, $c(x)$ are sufficient smooth functions satisfying $b(x) > 0$ and $c(x) > 0$ in $\overline{\Omega}$, and where the function $f(x)$ is sufficiently smooth and the numbers u_0, u_1 are given.

The multidimensional elliptic equation of second-order is further given as a boundary value problem:

$$Lu := -\nabla D \nabla u + b(x)^t \, \nabla u + c(x) \, u = f(x) \text{ in } x \in \Omega \subset \mathbb{R}^d \text{(B.92)}$$
$$\text{with } u(x) = u_0(x) \text{ on } \partial\Omega, \tag{B.93}$$

where $D \in \mathbb{R}^{d,+} \times \mathbb{R}^{d,+}$ is a matrix and $b : \Omega \to \mathbb{R}^{d,+}$ is sufficient smooth and also the function $f : \Omega \to \mathbb{R}$. The Dirichlet boundary is given with the function $u_0 : \partial\Omega \to \mathbb{R}$.

REMARK B.11 L denotes the second-order partial differential operator, where we have given the nondivergence free form. We also assume the symmetry condition $D = D^T$. ☐

In the next sections we briefly introduce the spatial discretization methods, which are used for the decomposition methods.

B.3.3 Finite Difference Method

The early development of numerical analysis of partial differential equations was dominated by finite difference methods, see [146].

Because of the simple approximate solutions at the point of a finite grid and the approximation of the derivatives by difference quotients, the finite difference method was attractive to many test examples, see [45].

For the numerical solution of Equation (B.92), we deal with $N + 1$ mesh-points $0 = x_0 < x_1 < \ldots < x_N = 1$ by setting $x_j = jh$, $j = 0, \ldots, N$, where $h = 1/N$. The approximation of $u(x_i)$ is U_i and we define the following finite difference as an approximation for the derivatives:

$$\frac{\partial u}{\partial x} = \lim_{h \to \infty} \frac{u(x_i) - u(x_{i-1})}{h}, \tag{B.94}$$

$$\partial^- U_i = \frac{U_i - U_{i-1}}{h}, \tag{B.95}$$

where $h = x_i - x_{i-1}$ and $\partial^- U_i$ is the backward difference of U_i.

Further, the forward difference is given as

$$\partial^+ U_i = \frac{U_{i+1} - U_i}{h}, \tag{B.96}$$

where h is the spatial grid step.

We can also generalize the finite differences to multidimensional applications on rectangular grids, see [146]. The idea behind this is to introduce the finite differences for the next dimensions and to adapt the difference quotients.

Here we do not concentrate on the multidimensional case of the finite difference methods but show the limited application of such methods.

Because of assumptions for the consistency to the smoothness of the solutions, the application is limited to sufficient smooth problems.

Example B.2

To apply only the second-order derivatives, we have the following assumptions to the solutions:

$$|\partial^- \partial^+ U_i - | \leq Ch^2 |u|_{C^4}, \tag{B.97}$$

where C^4 is the set of four-times continuous differentiable functions in Ω. ⬚

Next we will define the consistency and stability of the finite difference methods, which is important to have a convergent scheme.

Consistency and Stability

To have convergent methods, consistency and stability are necessary, see [147].

So for the convergence we need the consistency of the method which can be defined as in the following definition.

DEFINITION B.17 We have a consistency of order k, if we have

$$\max_i |L_h(R_h(u(x_i))) - R_h(L(u(x_i)))| \leq Ch^k, \tag{B.98}$$

where L_h is the discretized operator, L is the continuous operator, and the operator R_h maps from the continuous to the discrete points, so it is a restriction to the underlying grid points.

So the consistency denotes the quality of the approximation of the differential operator.

Further, the stability of the method is needed. We can define the stability as in Definition B.18.

DEFINITION B.18 We have a stable method, if we have

$$||u_h||_\infty \leq C||f_h||_\infty, \tag{B.99}$$

where u_h is the approximate solution and f_h is the approximated right-hand side.

A short proof can be given in the following.

PROOF

The stability can be shown by the boundedness of the Green's function, see [114] and [175]. The exact solution can be derived as

$$u(x) = \int_0^1 G(x,t) \, f(t) \, dt, \qquad (\text{B.100})$$

where $G(x,t)$ is the Green function to the second-order equation. □

REMARK B.12 The stability denotes the behavior that the defect $L_h u - L_h u_h$ is related to the error $u - u_h$. We could say $L_\infty - L_\infty$-stability or $L_1 - L_\infty$-stability. □

So the convergence of order k is given in the following definition.

DEFINITION B.19 The difference method is convergent of order k, if we have

$$\max_{i=1}^{N-1} |u(x_i) - U_i| \leq Ch^k, \qquad (\text{B.101})$$

where $u(x_i)$ is the exact solution at grid point x_i and U_i the approximated solution. h is the grid step and C is a constant independent of h.

REMARK B.13 The consistency order and the stability results in the convergence order. So the higher convergence order is obtained by higher consistency order. Therefore, the underlying solution has to be more smooth. For sufficient smooth problems, the finite difference methods are a well-known discretization method. □

Problems of the Finite Difference Methods

While the simple implementation and theoretical background are often the case for applying the finite difference as a first discretization method of a test example, there are several drawbacks that restrict the application to our model problems.

The general application of the difference methods is restricted, while the following problems appear:

- The delicate construction of the finite difference schemes for general grids

- Need of high regularity of the solution for the difference schemes

Therefore, we propose the finite element and finite volume methods for our applications of the decomposition methods. They are introduced in the following sections.

B.3.4 Finite Element Methods

The finite element methods were developed in the 1960s and 1970s by engineers, and in the 1970s and 1980s the mathematical theory behind the ideas was done, see [29], [44], and [187].

The basic ideas are the approximation of partial differential equations on the variational form of the boundary value problems and approximation of the piecewise polynomial function.

Because of the robustness of the method to general geometries and to weak derivatives, an application to arbitrary domains and less smooth solutions are allowed.

A well-known approach is the Galerkin's method, which consists in defining a similar problem, called a discrete problem, which is defined over a finite dimensional subspace V_h of the space V, see [44].

Then we have the three basic aspects of the finite element method:

1. Triangulation T_h.

2. The functions $v_h \in V_h$ are piecewise polynomials in the sense of each $K \in T_h$ in the sense that for each $K \in T_h$ the spaces $P_k = \{v_{h|K}; v_h \in V_h\}$ consist of polynomials.

3. There should exist a basis in the space V_h whose functions have small support.

For the illustration of the finite element methods, we deal with following boundary problem:

$$Lu := -(a(x)u')' + c(x)\,u = f(x)\,, \text{ in } x \in \Omega = (0,1), \quad \text{(B.102)}$$
$$\text{with } u(0) = u(1) = 0, \quad \text{(B.103)}$$

where $a(x)$, $c(x)$ are sufficient smooth functions satisfying $a(x) > 0$ and $c(x) > 0$ in $\overline{\Omega}$, and where the function $f \in L_2 = L_2(\Omega)$.

The variational formulation of this problem is to find $u \in H_0^1$:

$$a(u, \phi) = (f, \phi)\,, \ \forall \phi \in H_0^1, \quad \text{(B.104)}$$

where

$$a(v, w) = \int_\Omega (a\,v'\,w' + c\,v\,w)\,dx\,, \text{ and } (f, v) = \int_\Omega f v\,dx, \quad \text{(B.105)}$$

and the problem has a unique solution $u \in H^2$.

Therefore, the basic aspects of the discretization spaces, the error estimates, and the variational schemes are important and are discussed in the following subsections.

B.3.4.1 Sobolev Spaces

For the variational formulation, the application of weak derivates is important to have the solution spaces. We introduce the underlying spaces, Sobolev spaces, which is an application of the Lebesgue norms and spaces to include the weak derivatives.

We have the following definition of the weak derivatives.

DEFINITION B.20 Let k be a nonnegative integer, and let $f \in L^1_{loc}(\Omega)$. We assume the existence of weak derivatives $D^\alpha_w(f)$ for all $|\alpha| \leq k$. We define the Sobolev norm:

$$||f||_{W^k_p(\Omega)} := \left(\sum_{|\alpha| \leq k} ||D^\alpha_w f||^p_{L^p(\Omega)} \right)^{1/p}. \tag{B.106}$$

We define the Sobolev spaces via

$$W^k_p(\Omega) := \left\{ \sum_{|\alpha| \leq k} ||D^\alpha_w f||^p_{L^p(\Omega)} \right\}. \tag{B.107}$$

THEOREM B.3

The Sobolev space $W^k_p(\Omega)$ is a Banach space.

THEOREM B.4

Let Ω be any open set. Then $C^\infty(\Omega) \cup W^k_p(\Omega)$ is dense in $W^k_p(\Omega)$ for $p < \infty$.

We have the inclusion of Sobolev spaces.

PROPOSITION B.1

Let Ω be any open set. Then $C^\infty(\Omega) \cup W^k_p(\Omega)$ is dense in $W^k_p(\Omega)$ for $p < \infty$.

In the following we discuss the application of the weak formulation, as the variational principle, also known as Ritz-Galerkin approach, see [65] and [187].

B.3.4.2 Ritz-Galerkin Approach

The idea is to find an unique solution of the weak formulation within a (finite-dimensional) subspace.

Therefore, the unique representation of the linear functional is important. We formulate the Riesz Representation Theorem.

THEOREM B.5

*We can represent any continuous linear functional F on a Hilbert space H
as*

$$F(v) = (u, v), \tag{B.108}$$

for some $u \in H$, and we have

$$\|F\|_{H'} = \|u\|_H. \tag{B.109}$$

PROOF

     The uniqueness can be followed by the proof in [29].   ☐

Further we suppose the following conditions:

H.1. $(H, (\cdot, \cdot))$ is a Hilbert space.

H.2. V is a (closed) subspace of H.

H.3. $a(\cdot, \cdot)$ is a bounded, symmetric bilinear form that is coercive on V.

The Ritz-Galerkin approach is the following.
Given a finite-dimensional subspace $V_h \subset V$ and $F \in V'$, find $u_h \in V_h$ such
that

$$a(u_h, v) = F(v) \; \forall v \in V_h. \tag{B.110}$$

We can follow the uniqueness of our formulation B.110

THEOREM B.6

Under the conditions H.1.–H.3., there exists a unique u_h that solves (B.110).

PROOF

While $(V_h, a(\cdot, \cdot))$ is a Hilbert space and $F|_{V_h} \in V_h'$, we can apply the Riesz
Representation Theorem and obtain uniqueness.   ☐

In the following we discuss the error estimates.

B.3.4.3   Error Estimates

For the convergence of a Galerkin method we could use the Cea's lemma.

THEOREM B.7

*Let $a(\cdot, \cdot)$ be a continuous, V-elliptic bilinear form, and for all $f \in V'$
we could find an unique solution by the Riesz formulation. Then we could*

estimate

$$\|u - u_h\|_{H^1} \leq \frac{M}{\gamma} \inf_{v_h \in V_h} \|u - v_h\|_{H^1}, \tag{B.111}$$

where $V = H^1$ and H^1 is a Hilbert space.

PROOF

An unique solution for (B.150) is based on the Lax-Milgram lemma, so we could use the behavior also for $V_h \subset V = H^1$.

We have $a(u, v_h) = f(v_h)$, $\forall v_h \in V_h$, thus also $a(u - u_h, v_h) = 0$, $\forall v_h \in V_h$, and $a(u - u_h, u_h) = 0$.

Therefore $a(u - u_h, u - u_h) = a(u - u_h, u - v_h)$, $\forall v_h \in V_h$.

Then we have with continuity and V-ellipticity: $\gamma \|u - u_h\|^2 \leq M \|u - u_h\| \|u - v_h\|$. ◻

REMARK B.14 The partial spaces $V_h \subset V$ are chosen so that they are asymptotically dense in V.

So each element $u \in V$ has a good approximation by the elements $v \in V_h$, if $h > 0$ is sufficiently small and we have the convergence (B.150): $\lim_{h \to 0} \|u - u_h\| = 0$, and we have the approximation order: $\inf_{v_h \in V_h} \|u - v_h\|$.

The error boundary of the infimum is estimated by an interpolation operator Π_h:

$$\inf_{v \in V_h} \|u - v\| \leq \|u - \Pi_h u\| \tag{B.112}$$

with regularity we could estimate for special test spaces.

◻

An overview to the variational formulation with finite discretization methods is given in Figure B.2.

B.3.4.4 Construction and Design of Finite Element Methods

We deal with the approximation of the variational problem.
Find $u \in V$ such that

$$a(u, v) = F(v), \quad \forall v \in V, \tag{B.113}$$

where V is a Hilbert space, $F \in V'$ and $a(\cdot, \cdot)$ is a continuous, coercive bilinear form.

For such a problem, we can construct finite-dimensional subspaces $V_h \subset V$ in a systematic, practical way, see [44].

We deal with three questions:

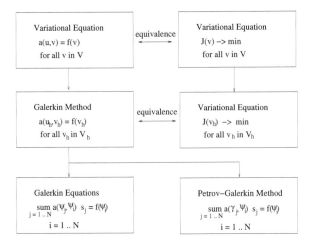

FIGURE B.2: Variational formulations with finite discretizations methods.

1. Simple trial and test space for approximating the operators and functions in the subinterval.

2. A good determination of the function in the subinterval.

3. A transition for the trial function to save the global behavior.

Construction Criteria for the Finite Element Methods
1) Triangulation of the continuous domain

   DEFINITION B.21 Let

   (i) $K \subset \mathbb{R}^n$ be a bounded closed set with nonempty interior and piecewise smooth boundary (the element domain),

   (ii) $\mathcal{P}$ be a finite dimensional space of functions on K (the space of shape functions), and

   (iii) $\mathcal{N} = \{N_1, N_2, \ldots, N_k\}$ be a basis for $\mathcal{P}'$ (the set of nodal variables).

Then $(K, \mathcal{P}, \mathcal{N})$ is called a finite element.

2) Definition of a test and trial space:
A linear test space is given as

$$\phi_i(x) = \left\{ \begin{array}{ll} \frac{1}{h_i}(x - x_i) & , \ x \in \Omega_i \\ \frac{1}{h_{i+1}}(x_{i+1} - x) & , \ x \in \Omega_{i+1} \\ 0 & \text{else} \end{array} \right\}, \tag{B.114}$$

for one element.

3) Local and global characteristics:

Here we deal with conform elements. The discrete problem is to embed into the continuous problem and we obtain: $V_h \subset V$.

For example $H^1(\Omega)$ and $H^2(\Omega)$ in the second-order elliptic problem and we obtain the following implication, see Lemma B.1.

LEMMA B.1

Let $z : \overline{\Omega} \rightarrow R$ be a function, with $z|_{\Omega_j} \in C^1(\overline{\Omega_j})$, with $i = 1, \ldots, N$. Then we have the implication

$$z \in C(\overline{\Omega}) \rightarrow z \in H^1(\Omega). \tag{B.115}$$

$\square$

REMARK B.15

The discretization errors can be classified into two groups:

1. Approximation error from the approximation of the operators and functions

2. Approximation error from the numerical integration

The first error is estimated by the consistency and stability.

The second error is an error of the approximate integration. Using polynomials we could construct exact integration formulas (e.g., Gauss quadrature formulas).

$\square$

Triangular Finite Elements

Let K be any triangle. Let $\mathcal{P}_k$ denote the set of all polynomials in two variables of degree less or equal k. We have the following dimensions for the elements.

The Lagrange element is given in the following example.

For $(k = 1)$, let $\mathcal{P} = \mathcal{P}_\infty$. Let $\mathcal{N}_1 = \{N_1, N_2, N_3\}$ $(dim\mathcal{P}_1 = 3)$ where

$$N_i(v) = v(z_i), \tag{B.116}$$

and z_1, z_2, z_3 are the vertices of K.

Further we can define the local interpolant for each finite element.

DEFINITION B.22 Given a finite element $(K, \mathcal{P}, \mathcal{N})$, let the set $\{\phi_i : 1 \leq i \leq k\} \subset \mathcal{P}$ be the dual basis to $\mathcal{N}$. If v is a function for which all $N_i \in \mathcal{N}$, $i = 1, \ldots, k$, are defined, then we define the local interpolant by

$$\mathcal{T}_K v := \sum_{i=1}^{k} N_i(v)\phi_i. \tag{B.117}$$

Example: Let K be any triangle.

$$T_K f = N_1(f)(1 - x - y) + N_2(f)x + N_3(f)y \qquad (\text{B.118})$$
$$= 1. \qquad (\text{B.119})$$

B.3.4.5 Theory of Conforming Finite Element Methods

The conforming theory of finite element methods is based on the Ritz method for symmetric and on the Galerkin method for the nonsymmetric bilinear forms.

Important is the embedding of the finite space into the infinite or continuous space: $V_h \subset V$.

We could formulate Theorem B.8.

THEOREM B.8

Because of $V_h \subset V$ and with the same scalar product $(\cdot, \cdot)$ we could define a Hilbert space and $a(\cdot, \cdot)$ has for V_h the same behavior as for V.

Then there exists an unique solution $u_h \in V_h$ with the sufficient optimality criteria:

$$a(u_h, v_h) = f(v_h) \text{ for all } v_h \in V_h. \qquad (\text{B.120})$$

This approximative treatment is used as the Ritz method. For unsymmetric bilinear forms the discrete variational formulation is called the Galerkin method.

Error Estimates for the Conforming Methods

For the convergence of a Galerkin method we could use the **Cea's lemma**.

THEOREM B.9

Let $a(\cdot, \cdot)$ be a continuous, V-elliptic bilinear form, such that for all $f \in V^*$ we could find a unique solution by the Riesz formulation. Then we could estimate

$$||u - u_h|| \leq \frac{M}{\gamma} \inf_{v_h \in V_h} ||u - v_h||. \qquad (\text{B.121})$$

PROOF

The unique solution for (B.150) is based on the Lax-Milgram lemma, so we could use the behavior also for the subspace $V_h \subset V$.

We have

$$a(u, v_h) = F(v_h) \; \forall \; v_h \in V_h , \qquad (\text{B.122})$$

so also,

$$a(u - u_h, v_h) = 0 \ \forall \ v_h \in V_h, \tag{B.123}$$

and

$$a(u - u_h, u_h) = 0. \tag{B.124}$$

Therefore, we have

$$a(u - u_h, u - u_h) = a(u - u_h, u - v_h) \ \forall \ v_h \in V_h. \tag{B.125}$$

Then we have with continuity and V-ellipticity

$$\gamma \, ||u - u_h||^2 \leq M \, ||u - u_h|| ||u - v_h||. \tag{B.126}$$

$\square$

Regularity for Solution and Embedding of Solutions to Strong Solutions

The idea is to embed the classical solution into the weak solutions with the boundary conditions.

REMARK B.16 The regularity of the solution and right-hand side is not enough, the regularity of the domain is also necessary. $\square$

The problem of the regularity can be shown in the following example.

Example B.3

We deal with a domain, including a discontinuous edge, see Figure B.3.

Discontinuous Edge

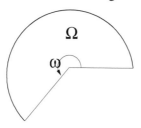

FIGURE B.3: Domain with Discontinuous Edge.

We have the analytical solution, see [27], given as

$$u(r, \phi) = r^{\pi/(2\omega)} \ \sin(\frac{\pi}{2\omega} \ \phi). \tag{B.127}$$

For example, with $\omega = \pi$, we have a singularity in $r \to 0$ such that $u \notin H^2(\Omega)$.

□

So we do not have a strong solution because of the discontinuity of the singular point $r \to 0$.

In the following we present the embedding to obtain continuity of the solutions.

THEOREM B.10

The embedding formula for the continuity of the solution (e.g., weak solutions are embedded in strong solutions), is given as
   Continuous embedding:
   We have $f \in H^k(\Omega)$ with $k > n/2$. Then we have for the solution,

$$u \in C^2(\overline{\Omega}) \cap H_0^1(\Omega), \tag{B.128}$$

and u is a classical solution.

PROOF

Because of $u \in H^{k+2} \cap H_0^1$, we have $u \in H^{k+1}(\Omega)$ and with the continuous embedding $W_2^{k+2}(\Omega) \to C^2(\overline{\Omega})$ for $k > n/2$ we get the assumption.

□

B.3.4.6 Interpolant (Reference Element), Theory of Conforming Elements

The interpolation property could be studied more simply by introducing a reference discretization.

LEMMA B.2

Given is a partial domain K and the reference element K' with an affine translation:

$$x = F(p)' = Bp + b \text{ for } p \in K'. \tag{B.129}$$

Then we have the following properties:

   i) $u \in H^l(K) \leftrightarrow v \in H^l(K')$,

   ii) $|v|_{l,K'} \leq c\|B\|(detB)^{-1/2}|u|_{l,K} \in H^l(K)$.

□

Example: quasi-uniform triangulation:

$$|v|_{r,K'} \leq c\{ \inf_{p \in K'} s(p)^{-1/2} h^r |u|_{r,K}. \tag{B.130}$$

Interpolation Error of the Conforming Theory
In the conforming theory the interpolation theory is important.
The idea is to find a "good" projection operator:

$$||u - u_h|| \leq c \inf_{v \in V_h} ||u - v|| \leq c||u - \Pi_h u||.$$

We have the following theorem:

THEOREM B.11

*We have the piecewise defined projector over the triangulation Z and for
polynomials of order k. For $r \leq k$ and regularity conditions for the reference
element we could denote*

$$||u - \Pi_Z u||_r \leq ch^{k+1-r}||u||_{k+1}, \tag{B.131}$$

where $h > 0$ assigns the fineness of the triangulation Z.

**Theory of the Conforming Elements to Nonlinear Boundary Value
Problems**
We could enlarge our theory to nonlinear boundary value problems. The
idea is to describe the monotone operators.

Important are the monotony and the start position of the iteration.

We could transfer the linear results to the nonlinear results, if we have the
following characterization:

- a) It exists a $\gamma > 0$ with: $(Bu - Bv, u - v) \geq \gamma||u - v||^2$ for all $u, v \in V$
  (we call the operator to be strong monotone).

- b) It exists a $M > 0$ with: $||Bu - Bv|| \leq M||u - v||$ for all $u, v \in V$.

We treat the abstract operator equation:

$$(Bu, v) = 0 \text{ for all } v \in V. \tag{B.132}$$

A generalization of the Lax-Milgram lemma is given in the following lemma
(idea: fixed point iteration).

LEMMA B.3
With respect to a) and b), the operator equation $Bu = 0$ has an
unique solution $u \in V$, which is the fixed point:

$$T_r v := v - rBv \text{ , for all } v \in V, \tag{B.133}$$

with $T_r : V \to V$ for the parameter values $r \in (0, \frac{2\gamma}{M^2})$.

□

In the next section we discuss the nonconforming theory of the finite element method, which can help to reduce the complexity of the formulation by an approximation.

B.3.4.7 Nonconforming Theory

The motivation for nonconforming elements is given by

1) The choice of the function spaces $V_h \subset V$ is too complex, for example with a differential equation of higher order, we have $V_h \not\subset V$.

2) The bilinear form and the linear functional are only approximately computable (e.g., with quadrature formulas).

3) Inhomogeneous, natural boundary conditions or curved boundaries of the domain Ω do not permit an exact projection of the boundary conditions.

The nonconforming finite element methods are finite element methods that do not directly discretize:

$$a(u, v) = f(v) \text{ for all } v \in V, \tag{B.134}$$

with $V_h \subset V$.

We can formulate our finite-dimensional problem as

$$a_h(u_h, v_h) = f_h(v_h) \text{ for all } v_h \in V_h, \tag{B.135}$$

where $a_h : V_h \times V_h \to \mathbb{R}$ a continuous bilinear form with V_h-ellipticity, see

$$\tilde{\gamma}\|v_h\|_h^2 \le a_h(v_h, v_h) \text{ for all } v_h \in V_h. \tag{B.136}$$

With the case $V_h \not\subset V$, the bilinear form a_h and the linear functional f_h are not certainly defined on V.

So we use the spaces Z_h and Z with $V \to Z$ and $V_h \to Z_h \to Z$ with $a_h(\cdot, \cdot)$ and $f_h(\cdot)$ being defined on $Z_h \times Z_h$ and Z_h.

Here we obtain the second Strang lemma.

LEMMA B.4
Second Lemma of Strang, see [186]:

$$\|u - u_h\| \le c \inf_{z_h \in V_h} \{\|u - z_h\| + \|f_h - a_h(z_h, \cdot)\|_{*,h}\}. \tag{B.137}$$

$\square$

Approximation of the Operators and Functionals

It is important to have the approximation of underlying integration. Our problem is given as

$$\forall u \in V, \ a(u, v) = f(v), \forall v \in V,$$
(B.138)

and the numerical approximation is given as:

$$\forall u_h \in V_h, \ a_h(u_h, v_h) = f(v_h), \forall v_h \in V_h.$$
(B.139)

We can deal with the following formulations of the elements :

1. Conforming elements: We have $V_h \subset V$ and the bilinear form a_h being V_h-elliptic.

2. Nonconforming elements: We have $V_h \not\subset V$ and the bilinear form a_h being V_h-elliptic.

Estimation of the conforming and nonconforming elements is given in the following lemma.

We have the general error estimation for the approximate operators and functionals.

LEMMA B.5

There holds

$$||u - u_h|| \le c \inf_{z_h \in V_h} \{||u - z_h|| + ||f_h - a_h(z_h, \cdot)||_{*, h}\}.$$

▯

REMARK B.17

- The choice of the nonconforming space Z is important to minimize the error of the numerical approximation.

- $V \hookrightarrow Z$, the continuity is satisfied for the solution of the equation, while the spaces are embedded in the space of the integration.

- $V_h \hookrightarrow Z_h \hookrightarrow Z$, also the discrete solution space is embedded into the integration space.

▯

Further, the approximation of the integration is given in the following lemma.

LEMMA B.6

First Lemma of Strang, see [186].

We have $V_h \subset V$ and the bilinear form a_h being V_h-elliptic, then we have a $c > 0$ such that

$$\|u - u_h\| \leq c \inf_{z_h \in V_h} \{\|u - z_h\| + \|a(z_h, \cdot) - a_h(z_h, \cdot)\|_{*,h}$$

$$+ \|f - f_h\|_{*,h}\} \tag{B.140}$$

is fulfilled.

$\square$

In the next subsection we present some examples of the elements.

B.3.4.8 Example: Conforming and Nonconforming Element for a Plate Equation

Based on the embedding theorem,

$$W^{j+m,p}(\Omega) \to C^j(\overline{\Omega}_0) \text{ if } mp > n, \tag{B.141}$$

we need very high polynomial degrees for higher-order operators.

For example, look at the plate equation:

$$\Delta^2 u = f \text{ in} \Omega, \tag{B.142}$$

$$u = \phi_1, \quad \frac{\partial u}{\partial n} = \phi_2 \text{ on } \Gamma. \tag{B.143}$$

For a conforming element for one dimension, we need the first derivation, thus the embedding should be in C^1, such that for $p = 2$ we have $j = 1$ and $m = 2$ and so we have H^2, which means we need third-order polynomials.

The element is given as $w(x) = c_1 + c_2 x + c_3 x^2$.

But for $n = 2$ we need H^3 and thus a 16-parameter element as minimum (polynomial order 3).

A nonconforming element could be of a simpler reconstruction and we could reduce the order, so we choose $w(x) = c_1 + c_2 x$, and we are only in C^1.

B.3.4.9 Extension to Isoparametric Elements and Curved Domains

Here the idea of this discretization method is to obtain improved results at the curved boundary.

The affine transformation from an element to this reference element is done and this is added to the interpolation error.

We can formulate the following lemma.

LEMMA B.7

We have the following estimation of our interpolation error:

$$|v|_{m,p,K'} \leq c \|B\|^m |det(B)|^{-1/p} |v|_{m,p,K}\}, \tag{B.144}$$

where $x = Bx' + b$ is the iso-parametric mapping.

☐

The isoparametric mapping is presented in Figure B.4.

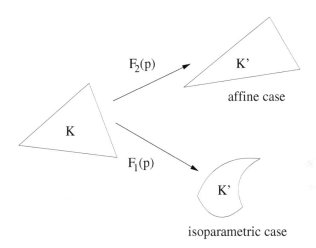

FIGURE B.4: Isoparametric mapping.

REMARK B.18 Because of the mapping for the elements, the application to the isoparametric elements is widespread in the engineering problems, see [20]. Mathematically, one can deal with the transformation and treat the approximations with the unit element, see [46]. ☐

B.3.5 Finite Volume Methods

Finite volume methods are mostly applied in differential equations with divergenceform, so, for example, parabolic differential equations with conservation terms. One can see the theory of the finite volume methods with ideas of the finite difference as also of the finite element methods. So the formulation can be done as a generalized difference method or as a variant of the finite element methods.

The fundamental works in this field are done for example in [113] and [121].

In the following we introduce the finite volume methods.

B.3.5.1 Motivation to Finite Volume Methods

The finite volume method is aligned to the finite difference method for unstructured grids.

Finite volume methods are also Petrov-Galerkin finite element methods.

Properties

- Conservativity: inner flux terms are added to be zero.

- Discretization of conservation laws and balance equations.

- Dual boxes allow simple discretizations.

- Theory of finite elements could be used for error estimates.

- Barycentric cells are needed for the barycentric method in $\mathbb{R}^n$.

Finite element methods are not conservative, which is a problem in nonlinear and discontinuous parameters.

B.3.5.2 Model Equation: Convection-Diffusion Equation

We discuss in the following parabolic differential equation of second order. We have for the linear elliptic equation the form

$$Lu := -\nabla \cdot (A\nabla u - v\, u) + ru = f, \qquad (B.145)$$

where $A : \Omega \to \mathbb{R}^{d,d}$, $v : \Omega \to \mathbb{R}^d$ and $r, f : \Omega \to \mathbb{R}$.

The parabolic equation is given as

$$\frac{\partial u}{\partial t} + Lu = f, \qquad (B.146)$$

where we have the initial and boundary conditions given as

$$u(x,0) = u_0(x)\,, x \in \Omega\,, \qquad (B.147)$$
$$u(x,t) = \bar{u}(x,t)\,, (x,t) \in \partial\Omega \times (0,T), \qquad (B.148)$$

B.3.5.3 Formulation of the Finite Volume Method

In the following we present the formulation of the finite element method and also finite volume method to see the differences.

First formulation (FEM formulation or Galerkin formulation), we have $u \in H_0^1(\Omega)$,

$$\int_\Omega \nabla v \cdot A\nabla u\, dx - \int_\Omega ub \cdot \nabla v dx + \int_\Omega cuv dx \qquad (B.149)$$
$$= \int_\Omega fv\, dx\,, \forall v \in H_0^1(\Omega).$$

The second formulation is the Petrov-Galerkin method, which is the finite volume formulation, we have $u \in H^2(\Omega) \cap H_0^1(\Omega)$,

$$\sum_{B \in \mathcal{B}} \left(\int_B \nabla \bar{v}(A\nabla u - bu)dx + \int_B cu\bar{v}dx - \int_{\partial B} \bar{v}A\nabla u \, d\sigma \right) \quad \text{(B.150)}$$

$$+ \int_B bu\bar{v}dx = \int_\Omega f\bar{v}dx \, , \, \forall \bar{v} \in H_0^1(B),$$

The idea is to have two different spaces, for example the primary and the dual grid. For the dual grid we could choose a simpler space, for example, a box space:

$$\bar{v} = \begin{cases} 1 \, , \, x \in B \\ 0 \, , \, \text{else} \end{cases} . \quad \text{(B.151)}$$

The primary space could be chosen in a better space, for example, the P_1 or P_2 space (Lagrange space) of the finite element methods.

Geometrical Construction of the Finite Volume Method

In the geometrical construction of the method, the underlying dual cells are important. In the construction we have the local mass conservation, we obtain simple test functions (box functions), and we can apply the construction to unstructured grids (adaptive grids).

The first example is the standard first-order method based on the barycentric meshes, see Figure B.5.

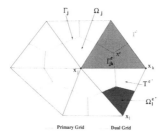

FIGURE B.5: The primary and dual mesh of the finite volume method.

The construction is given with the T^e elements, $e = 1, \ldots, E$ number of elements and the finite volume cells or dual cells Ω_j, $j = 1, \ldots, N$ number of nodes. The dual cells are constructed by the barycentre of the elements and the middle-point of each side. We obtain the tessellation of the dual cells, see also Figure B.5.

A second example is a higher-order finite volume discretization by P^2 Lagrangian Elements, see Figure B.6.

We have a finer discretization with the dual cells to obtain the conditions for the P^2 elements. The trial space is given with P^2 elements, where the test space is given with the constant P^0 elements, see [14].

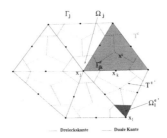

FIGURE B.6: The primary and dual mesh of the higher-order finite volume method.

For the dual grid $\mathcal{B}$, where the unknown solution u the nodes, we can construct a first-order method as follows.

DEFINITION B.23 We have $\Omega \subset \mathbb{R}^n$ being a polyhedral Lipschitz domain and $\mathcal{T}_h$ a consistent triangulation of Ω with the edge points $x_1^h, \ldots, x_{N_h}^h$. The set $\mathcal{B}_h = \{B_1^h, \ldots, B_{N_h}^h\}$ with the closed subsets $B_j^h \subset \overline{\Omega}$ is a dual box grid for $\mathcal{T}_h$, if we have the following conditions.

For the construction of the finite volume methods, we have to fulfill the following conditions:

1. Regularity Conditions

DEFINITION B.24 We have $\Omega \subset \mathbb{R}^n$ being a polyhedral lipschitzian domain and $\mathcal{T}_h$ a consistent triangulation of Ω. A dual box grid $\mathcal{B} = \{B_1^h, \ldots, B_{N_h}^h\}$ for $\mathcal{T}_h$ fulfills the regularity condition, if

(R1) All box parts $B_j^h = T, 1 \le j \le N_h, T \in \mathcal{T}_h$ are lipschitzian sets, cf. (a set $M \subset \mathbb{R}^n$ is a lipschitzian set, if $int(M)$ is a lipschitzian set and $\overline{M} = \overline{int(M)}$). $\mathcal{B}$ fulfills the second regularity condition (R2), if

(R2) $S_{j,i_j,T}(T) \cap S_{i_j,T}(T) = \emptyset$ is fulfilled for all $1 \le j \le N_h$ and for all elements $T \in \mathcal{T}_j^h$.

2. Equilibrium Conditions

DEFINITION B.25 We have $\Omega \subset \mathbb{R}^n$ being a polyhedral lipschitzian domain and $\mathcal{T}_h$ a consistent triangulation of Ω. A dual box grid $\mathcal{B} = \{B_1^h, \ldots, B_{N_h}^h\}$ for $\mathcal{T}_h$ fulfills the equilibrium condition, if

(G1)

$$vol(B_j^h \cap T) = \frac{vol(T)}{n+1}. \tag{B.152}$$

It fulfills the second equilibrium condition if we have for all boxes $B_j^h \in \mathcal{B}_h$ an element $T \in \mathcal{T}_j^h$ and for all indices $0 \leq k \leq n$ with $k \neq i_{j,T}$ the condition

(G2)

$$vol(S_{j,k}(T) = vol(B_j^h \cap S_k(T)) = \frac{vol(S_k(T)}{n}. \tag{B.153}$$

In the following we can derive error estimates for the finite volume method.

Global Error Estimates
The error estimate in the H^1-norm is given by

THEOREM B.12

$$\|u - u_h\|_{1,2} \leq \frac{CM_0}{\alpha}(h^s\|u\|_{1+s,2} + h\|f\|_{0,2}), \tag{B.154}$$

whereby the constant $C = C(\Omega, \delta, s)$ is independent from $h, u,$ and f.

PROOF

We use the first Strang lemma and get the abstract error estimates. $\Box$

Regularity for the Grid, Simplices
It is important for the hierarchical refinement that the conditions for the finite volume methods have to be fulfilled.

Therefore, we have to deal with the consistent refinement that is introduced in the following.
Topology of the simplices: $T \subset \mathbb{R}^n$,

$$vol\, T_n = \frac{1}{n!}. \tag{B.155}$$

Bases of simplices:

$$\chi_{\Pi(K)} = \sum_{j=k}^N \lambda_j, \tag{B.156}$$

A consistent triangulation is given with the Kuhn-Theorem, see [25].

The regular or irregular refinement is given as follows:

1) Regular refinement: 2^n simplices for each refinement, see Figure B.7.

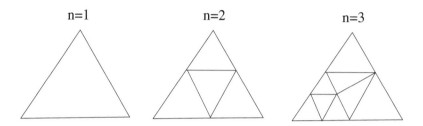

n=1   n=2   n=3

FIGURE B.7:   Regular refinement of simplices.

2) Irregular refinement: less than 2^n: For this case we have to use the ideas of Brower's fixed point theorem. Therefore, a hierarchical refinement is possible and the application of multigrid methods is possible.

B.3.5.4   Higher-Order Finite Volume Methods: Reconstruction Method and Embedding Methods

We discuss an improved second-order discretization method for the linear second-order parabolic equation by combining analytical and numerical solutions. Such methods are first discussed as Godunov's scheme, see [110] and [150], and use analytical solutions to solve the one-dimensional convection-reaction equation. We can generalize the second-order methods for discontinuous solutions because of the analytical test functions. One-dimensional solutions are used in the higher-dimensional solution of the numerical method.

The method is based on the flux-based characteristic methods and is an attractive alternative to the classical higher-order TVD-methods, see [119]. The analytical solution can be derived by using the Laplace transformation to reduce the equation to an ordinary differential equation. With general initial conditions (e.g., spline functions), the Laplace transformation is accomplished with the help of numerical methods. The proposed discretization method skips the classical error between the convection and reaction equation by using the operator-splitting method.

Embedding of Analytical Methods in Various Discretization Methods In this section we focus more on the mixing of the methods described in the first chapters. The overview for the different methods is given and the coupling between the methods is done.

The basic discretization methods as finite element, finite volume methods are done with polynomial spaces. These spaces could fit with results for smooth solutions, but a mixing for changes in the type of the equations is often problematic. For these problems, numerical artifacts are known as oscillations, numerical diffusion, and so forth, and are described in [119] and [143], especially for conservation laws.

The improvement for this new result came from the idea to stabilize the polynomial base functions.

In this section we discuss the embedding of analytical solutions for a convection-diffusion-reaction equation.

The analytical results for one, two, and three dimensions for linear convection-diffusion-reaction equations are derived in [62], [84], [106], and [129].

In the following, we present the embedded reaction equation to the convection equation.

Improved Discretization Methods via Exact Transport and Reaction on the Characteristics

The scalar equation is given by

$$\partial_t R\, c + \nabla \cdot \mathbf{v}\, c = 0.0, \qquad (B.157)$$

where the initial conditions are $c(x, t^0) = c^0(x)$.

The spatial integration plus the theorem of Gauss for the derivatives give

$$\int_{\Omega_j} \partial_t(R\, c)\, dx = -\int_{\Omega_j} \nabla \cdot (\mathbf{v}\, c)\, dx = -\int_{\Gamma_j} \mathbf{n} \cdot (\mathbf{v}\, c)\, d\gamma, \qquad (B.158)$$

where Ω_j is the j-th cell and $v_{jk} = \mathbf{n}_{jk} \cdot \int_{\Gamma_{jk}} \mathbf{v}(\gamma)\, d\gamma$.

$$|\Omega_j|(R(c_j^{n+1}) - R(c_j^n)) = -\tau^n \sum_{k \in out(j)} v_{jk}\tilde{c}_{jk}^n + \tau^n \sum_{l \in in(j)} v_{lj}\tilde{c}_{lj}^n. \qquad (B.159)$$

The discretization scheme with the mass notation is

$$m_j^{n+1} - m_j^n = -\sum_{k \in out(j)} m_{jk}^n + \sum_{l \in in(j)} m_{lj}^n, \qquad (B.160)$$

with

$$m_j^n = V_j\, R\, c_j(t^n)\,, \quad m_{jk}^n = \tau\, \tilde{c}_{jk}^n v_{jk}, \qquad (B.161)$$

with the limitation to satisfy the monotonicity (local min-max property). We use the reconstruction of the linear test function:

$$c_{jk}^n = c_j^n + \nabla c_j^n (x_{jk} - x_j). \qquad (B.162)$$

Limiters (slope and flux limiter):

$$\min_{k \in in(i)} \{c_i^n, c_k^n\} \le c_{jk}^n \le \max_{k \in in(i)} \{c_i^n, c_k^n\} \, , \, j \in out(i) \, , \qquad (B.163)$$

with limited value $\hat{c}_{jk}^n$,

$$\tilde{c}_{jk}^n = \hat{c}_{jk}^n + \frac{\tau}{\tau_j}(c_j^n - \hat{c}_{jk}^n) \, , \, \tau_j = \frac{V_j}{\mathbf{v}_j} \, , \qquad (B.164)$$

$$\mathbf{v}_j = \sum_{k \in out(j)} v_{jk} \, , \, v_{jk} = \mathbf{n}_{jk} \cdot \int_{\Gamma_{jk}} \mathbf{v}(\gamma) \, d\gamma \, . \qquad (B.165)$$

Modified Discretization with Finite Volume Methods of Higher-Order and Embedded Analytical Solutions

We now apply a Godunov's method for the discretization. We reduce the equation to a one-dimensional problem, solve the equation exactly, and transform the one-dimensional mass to the multidimensional equation. Therefore, we only get an error in the spatial approximation, as given for the higher-order discretization, and hence we can skip the time splitting error.

The equation for the discretization is given by

$$\partial_t u_i + \nabla \cdot \mathbf{v}_i \, u_i = -\lambda_i u_i + \lambda_{i-1} u_{i-1}, \qquad (B.166)$$
$$i = 1, \ldots, m \, . \qquad (B.167)$$

The velocity vector $\mathbf{v}$ is divided by R_i, and m is the number of concentrations. The initial conditions are given by $u_1^0 = u_1(x,0)$, $u_i^0 = 0$ for $i = 2, \ldots, m$. The boundary conditions are trivial conditions (i.e., $u_i = 0$ for $i = 1, \ldots, m$).

We first calculate the maximal time-step for cell j and for concentration i using the total outflow fluxes:

$$\tau_{i,j} = \frac{V_j \, R_i}{\mathbf{v}_j} \, , \quad \mathbf{v}_j = \sum_{j \in out(i)} v_{i,j}. \qquad (B.168)$$

We get the restricted time-step with the local time steps of cells and their components:

$$\tau^n \le \min_{\substack{i=1,\ldots,m \\ j=1,\ldots,I}} \tau_{i,j}. \qquad (B.169)$$

The velocity of the discrete equation is given by

$$v_{i,j} = \frac{1}{\tau_{i,j}} \, . \qquad (B.170)$$

We calculate the analytical solution of the mass with Equation (B.160) with

$$m_{ij,out}^n = m_{i,out}(a, b, \tau^n, v_{1j}, \ldots, v_{ij}, R_1, \ldots, R_i, \lambda_1, \ldots, \lambda_i) \, , \qquad (B.171)$$
$$m_{ij,rest}^n = m_{i,j}^n \, f(\tau^n, v_{1j}, \ldots, v_{ij}, R_1, \ldots, R_i, \lambda_1, \ldots, \lambda_i) \, , \qquad (B.172)$$

where the parameters are $a = V_j R_i (c^n_{i,jk} - c^n_{i,jk'})$, $b = V_j R_i c^n_{i,jk'}$, and $m^n_{i,j} = V_j R_i u^n_i$. The linear impulse in the finite volume cell is $c^n_{i,jk'}$ for the concentration at the inflow and $c^n_{i,jk}$ for the concentration at the outflow boundary of cell j.

The discretization with the embedded analytical mass is calculated by

$$m^{n+1}_{i,j} - m^n_{i,rest} = - \sum_{k \in out(j)} \frac{v_{jk}}{v_j} m^n_{i,jk,out} + \sum_{l \in in(j)} \frac{v_{lj}}{v_l} m^n_{i,lj,out} , \text{(B.173)}$$

where $\frac{v_{jk}}{v_j}$ is the retransformation of the total mass $m_{i,jk,out}$ into the partial mass $m_{i,jk}$. The mass in the next time-step is $m^{n+1}_{i,j} = V_j\, c^{n+1}_i$, in the old time-step it is the rest mass for the concentration i. The proof can be found in [81].

REMARK B.19 In the embedding methods, we assume to have the one-dimensional analytical or cheap by solving a simple ODE system. Based on this we can decouple the discretization methods to cheap parts (e.g., the reaction parts) and noncheap parts, the complicated spatial parts (e.g., elliptic parts). That knowledge can save memory and computational time in the discretization methods. ☐

Advanced Discretization Methods, Integration of Decomposing Methods

Often the standard methods are too general to obtain the needed accuracy. Therefore, new modified methods were developed to gain higher-order results. In Table B.1 we present such methods.

TABLE B.1: Modifications for the standard methods.

Discretization Method	Modification for the Method
Finite Volume	Reconstruction Method, see [14]
Finite Element	Enriched Methods, see [15], [123] (e.g., Bubble Functions)
Discontinuous Galerkin	Embedded Test Functions (e.g., One-Dimensional Solutions), see [85]

Especially for the discretization methods for parabolic differential equations, there exist several methods, for example,

1. Characteristic methods (exact transport and reaction): Test functions (linear or constant) are exact transported. Only approximation error

for the initial condition and splitting error in multiple dimensions, see [177].

2. Locally improved test functions: New improved test space for the finite volume or DG methods, locally exact solutions, see [85].

3. Locally improved trial functions to improve the discretization of critical terms. Idea: skip the critical terms via analytical solutions, see [84].

B.3.5.5 Extended Discretization Methods : Discontinuous Galerkin Methods

- Local mass conservation

- Higher-order methods (test functions with higher polynomials)

- Application for unstructured grids (adaptive grids)

- One grid (primary grid)

The direction of the element normal vector is given as in Figure B.8.

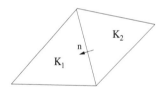

FIGURE B.8: Definition of the element normal vector.

- Triangulation $\mathcal{K}_h$ for $h > 0$ for the domain Ω

- Broken Sobolev space by

$$H^l(\mathcal{K}_h) = \{v \in L^2(\Omega) : v|_K \in H^l(K) \ \forall K \in \mathcal{K}_h\}. \quad (B.174)$$

Notation for Discontinuous Galerkin Methods

- Triangulation $\mathcal{K}_h$ for $h > 0$ for the domain Ω.

- Subdomain $K \in \mathcal{K}_h$ is a Lipschitz boundary.

- $\mathcal{E}_h^i$ of all interior boundaries e of $\mathcal{K}_h$.

- $\mathcal{E}_h^b$ of all exterior boundaries e of $\Gamma = \partial \Omega$.

We then have the jumps across the edge $e = \partial K_1 \cap \partial K_2$:

$$[v] = (v|_{K_2})|_e - (v|_{K_1})|_e . \tag{B.175}$$

We also have the averages on the interfaces:

$$\{v\} = \frac{(v|_{K_2})|_e + (v|_{K_1})|_e}{2} . \tag{B.176}$$

Mixed Formulation for the Convection-Diffusion-Reaction Equation

The solution is given by $u(x,t) \in C^2(\Omega) \times C^1([0,T])$ and $\underline{p}(x,t) \in (C^2(\Omega) \times C^1([0,T]))^d$ for the classical formulation, see also [85]:

$$\partial_t R u + \nabla \cdot \underline{v}\, u - \nabla \cdot a^{1/2}\, \underline{p} + R\lambda\, c = f \ , \ \text{in } \Omega \times [0,T], \tag{B.177}$$

$$-a^{1/2}\, \nabla u + \underline{p} = 0 \ , \text{in } \Omega \times [0,T], \tag{B.178}$$

$$u = g_1 \ , \ \text{on } \Gamma_1 \times [0,T] \ ,$$

$$(\underline{v}\, u - a^{1/2}\, \nabla u) \cdot \underline{n} = g_2 \ , \ \text{on } \Gamma_2 \times [0,T] \ ,$$

$$u(0) = u_0 \ , \ \text{in } \Omega.$$

The inner product $L^2(S)$ is denoted as $(\cdot,\cdot)_S$, and for $S = \Omega$ we skip the S.

Variational Formulation for the Mixed Equations

For the mixed methods we have to find the unknowns $\underline{p} \in L_2((H^1(\mathcal{K}_h))^d, [0,T])$ and $u \in L_2(H^1(\mathcal{K}_h), [0,T])$ as follows:

$$(R\, \partial_t u, \phi) - \sum_{K \in \mathcal{K}_h} (u, \underline{v} \cdot \nabla\phi)_K + \sum_{e \in \mathcal{E}_h} (h_{conv}(u), [\phi])_e$$

$$+ \sum_{K \in \mathcal{K}_h} < a^{1/2}\, \underline{p}, \nabla\phi >_K + \sum_{e \in \mathcal{E}_h} (h_{diff}(u_h), [\phi])_e + (R\, \lambda u, \phi)$$

$$= \sum_{e \in \mathcal{E}_h^D} (g_2, \phi)_e + (f, \phi) \ , \ \phi \in H^1(\mathcal{K}_h), \tag{B.179}$$

$$< \underline{p}, \underline{\chi} > + \sum_{K \in \mathcal{K}_h} (u, \nabla \cdot a^{1/2}\, \underline{\chi})_K + \sum_{e \in \mathcal{E}_h} (h_{diff}(\underline{p}_h), [a^{1/2}\, \underline{\chi} \cdot \underline{n}])_e$$

$$= \sum_{e \in \mathcal{E}_h^D} (g_1, \underline{\chi} \cdot \underline{n})_e \ , \ \underline{\chi} \in (H^1(\mathcal{K}_h))^d. \tag{B.180}$$

We have the convective fluxes:

$$\hat{h}_{conv}(u_h) = \begin{cases} \{u_h\, \underline{v}\, \underline{n}\} & \text{central differences} \\ \{u_h\, \underline{v}\, \underline{n}\} - \frac{|\underline{v}\, \underline{n}|}{2} [u_h] & \text{upwind} \end{cases} , \tag{B.181}$$

$$\hat{h}_{diff}(\underline{w}_h) = (a^{1/2}\, \underline{n}\, \{u_h\}, \{a^{1/2}\, \underline{p}_h \cdot \underline{n}\})^t + C_{diff}[(u_h, \underline{p}_h)^t], \tag{B.182}$$

where the flux matrix C_{diff} is given as

$$C_{diff} = \begin{pmatrix} 0 & -c_{1,2} & \cdots & -c_{1,d+1} \\ c_{1,2} & 0 & \cdots & 0 \\ \vdots & \vdots & \cdots & \vdots \\ c_{1,d+1} & 0 & \cdots & 0 \end{pmatrix} = \begin{pmatrix} 0 & -\underline{c}^T \\ \underline{c} & 0 \end{pmatrix},$$

where $\underline{c} = (c_{1,2}, \ldots, c_{1,d+1})^T$ and $c_{1,i} = c_{1,i}((\underline{w}_h|_{K_2})|_e, (\underline{w}_h|_{K_1})|_e)$ are locally Lipschitz.

Stability of the Mixed Form

We have the stability for the full-discrete form with the solutions $u_h \in V$ and $\underline{p}_h \in W$, such that

$$R \frac{1}{2} ||u_h(T)||^2_{L^2(\Omega)} + \int_0^T \Theta_C(\underline{w}_h, \underline{w}_h) \, dt + \int_0^T ||\underline{p}_h||^2_{(L^2(\Omega))^d} \, dt$$

$$+ \int_0^T \frac{R \lambda}{2} ||u_h||^2_{L^2(\Omega)} \, dt$$

$$\leq \frac{1}{2} ||u_h(0)||^2_{L^2(\Omega)} + \int_0^T \frac{1}{2 R \lambda} ||f||^2_{L^2(\Omega)} dt, \quad \text{(B.183)}$$

where $\Theta_C(\underline{w}_h, \underline{w}_h) = \sum_{e \in \mathcal{E}_h} < [\underline{w}_h], \, C \, [\underline{w}_h] >_e,$

and

$$C = \begin{pmatrix} c_{1,1} & -\underline{c}^T \\ \underline{c} & 0 \end{pmatrix} \quad \text{(B.184)}$$

is the flux matrix.

Design of New Test Functions with Local Analytical Solutions

For the adjoint problem of the convection-reaction terms, we get

$$- \sum_{K \in \mathcal{K}_h} (u_h, \underline{v} \cdot \nabla \phi)_K + (u_h, R \lambda \phi) \quad \text{(B.185)}$$

$$= \sum_{K \in \mathcal{K}_h} (u_h, -\underline{v} \cdot \nabla \phi + R \lambda \phi) = 0,$$

where $u_h, \phi \in V_h$, and we solve the adjoint local equation for the convection reaction in space:

$$-\underline{v} \cdot \nabla \phi + R \lambda \phi = 0, \quad \text{(B.186)}$$

where the initial condition is $\phi(0) = \phi_0$.

Embed Local Analytical Solutions to Test Functions of the Discontinuous Galerkin Methods

The motivation for the new test spaces came from the idea to improve the local behavior of the test functions. Standard test functions like polynomials do not respect the local character.

To have a local behavior of the solutions we use the ideas of the adjoint problems. They are done in the ELLAM schemes [177]. We use these ideas for the space terms and solve the locally adjoint problems for the spatial dimensions.

We concentrate on the convection-reaction equation, given as

$$- \sum_{K \in \mathcal{K}_h} (u_h, \underline{v} \cdot \nabla \phi)_K + (u_h, R \, \lambda \phi) \tag{B.187}$$

$$= \sum_{K \in \mathcal{K}_h} (u_h, -\underline{v} \cdot \nabla \phi + R \, \lambda \, \phi) = 0, \tag{B.188}$$

where $u_h, \phi \in V_h$, and solve the adjoint local equation for the convection reaction in space:

$$-\underline{v} \cdot \nabla \phi + R \, \lambda \, \phi = 0 \,, \tag{B.189}$$

where the initial condition is $\phi(0) = \phi_0$. We derive the local solution of the equation (B.189) the next section.

Local Test Functions in Space for One Dimension

We derive the one-dimensional solutions for the local convection-reaction equation, given as adjoint problem:

$$-v \, \partial_x \phi + R \, \lambda \, \phi = 0 \,, \tag{B.190}$$

whereby $\phi(0) = a_0$ is a constant.

The equation (B.190) is solved exactly and the solution is denoted with respect to the velocity $v \in \mathbb{R}$ and $v \neq 0$:

$$\phi_{anal,i}(x) = a_0 \begin{cases} \exp(-\beta \, (x_{i+1/2} - x)) \; v > 0 \\ \\ \exp(-\beta \, (x - x_{i-1/2})) \; v < 0 \end{cases} , \tag{B.191}$$

where $\beta = \frac{R \, \lambda}{|v|}$,

$$\phi_{new,i} = \phi_{anal,i}(x) \,, \tag{B.192}$$

where $x_{i-1/2} < x < x_{i+1/2}$.

The local solution is applied as an analytical weight with $0 \le \phi_{anal,i}(x) \le 1$. For this test function we have one degree of freedom, so that we could use only a constant initial condition.

For linear initial conditions we use the analytical test function and multiply it with the standard test function of first-order.

Thus, we have

$$\phi_{new,i}\phi_{stand,i}(x)\ \phi_{anal,i}(x)\ , \qquad\qquad (B.193)$$

whereby the standard test functions are given as polynomial functions $\phi_{stand}(x) = \{1, x, x^2, \ldots\}$.

The test functions are used for the cases $|v| >> \lambda \geq 0$, for the case $v = 0$ we use the standard test functions, cf. [48].

So we improve it for constant cell impulses. To present the test functions we have Figures B.9 and B.10 for two extreme cases: $\lambda << v$ and for $\lambda = v$.

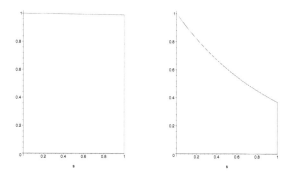

FIGURE B.9: Local test functions constructed with analytical solution and constant initial condition.

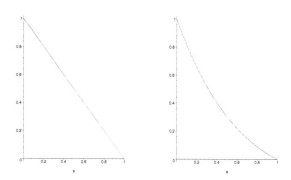

FIGURE B.10: Local test functions constructed with analytical solution and linear initial condition.

REMARK B.20 The local discontinuous Galerkin methods can be improved by local analytical solutions. Like the ELLAM schemes, we designed the test functions based on the dual formulation. That investigation can save memory and computational time in the discretization methods. ⬜

References

[1] M. Abramowitz and I.A. Stegun. *Handbook of Mathematical Functions.* Dover Publication, New York, 1970.

[2] G. Akrivis, M. Crouzeis, and Ch. Makridakis. *Implicit-explicit multistep methods for quasilinear parabolic equations.* Numerische Mathematik, 82:521–541, 1999.

[3] I. Alonso-Mallo, B. Cano, and J.C. Jorge. *Spectral-fractional step Runge-Kutta discretisations for initial boundary value problems with time dependent boundary conditions.* Mathematics of Computation, 73(248):1801–1825, 2004.

[4] Z. S. Alterman and A. Rotenberg. *Seismic waves in a quarter plane.* Bulletin of the Seismological Society of America, 59:347–368, 1969

[5] D.N. Arnold, F. Brezzi, B. Cockburn, and L.D. Marini. *Unified analysis of discontinuous Galerkin methods for elliptic problems.* SINUM Society for Industrial and Applied Mathematics, 39(5):1749–1779, 2002.

[6] U.M. Ascher, S.J. Ruuth, and B.T.R. Wetton. *Implicit-explicit methods for time-dependent partial differential equations.* SIAM Journal on Numerical Analysis, 32(3): 797–823, 1995.

[7] U.M. Ascher, S.J. Ruuth, and R.J. Spiteri. *Implicit-explicit Runge-Kutta methods for time-dependent partial differential equations.* Appl. Num. Math., 25(2–3):151–167, 1997.

[8] W. Balser and J. Mozo-Fernandez. *Multisummability of formal solutions of singular perturbation problems.* J. Differential Equations, 183(2): 526–545, 2002.

[9] W. Balser, A. Duval, and St. Malek, *Summability of formal solutions for abstract Cauchy problems and related convolution equations.* Manuscript, November 2006.

[10] W. Bangerth and R. Rannacher. *Adaptive finite element methods for differential equations.* Lectures in Mathematics, ETH Zürich, Birkhäuser Verlag, Basel-Boston-Berlin, 2003.

[11] V. Barbu. *Nonlinear Semigroups and Differential Equations in Banach Spaces.* Editura Academiei, Bucuresti, and Noordhoff, Leyden, 1976.

[12] V. Barbu. *Analysis and Control of Nonlinear Infinite Dimensional Systems.* Acad. Press, Boston, 1993.

[13] V. Barbu. *Partial Differential Equations and Boundary Value Problems.* Mathematics and its Applications, Kluwer Academic Publishers, Dordrecht, Boston, London, 1998.

[14] T. Barth and M. Ohlberger. *Finite volume methods: Foundation and analysis.* Encyclopedia of Computational Mechanics, editors: E. Stein, R. Borst, T.J.R. Hughes, West Sussex, England, John Wiley and Sons, Vol.1., chapter 15, 2004.

[15] I. Babuska, U. Banerjee, and J.E. Osborn. *Survey of meshless and generalized finite element methods: A unified approach.* Acta Numerica, Cambridge University Press, 12:1–125, 2003.

[16] D.A. Barry, C.T. Miller, and P.J. Culligan-Hensley. *Temporal discretization errors in non-iterative split-operator approaches to solving chemical reaction/groundwater transport models.* Journal of Contaminant Hydrology, 22: 1–17, 1996.

[17] P. Bastian. *Parallele adaptive Mehrgitterverfahren.* PhD thesis, University of Heidelberg, Germany, 1994.

[18] P. Bastian, K. Birken, K. Eckstein, K. Johannsen, S. Lang, N. Neuss, and H. Rentz-Reichert. *UG-a flexible software toolbox for solving partial differential equations.* Computing and Visualization in Science, 1(1):27–40, 1997.

[19] S. Bartels and A. Prohl. *Convergence of an implicit finite element method for the Landau-Lifschitz-Gilbert equation.* SIAM J. Numer. Anal., 44(4):1405–1419, 2006.

[20] K.J. Bathe. *Finite Element Procedures.* Prentice Hall, New Jersey, 1996.

[21] P. Bauer, S. Attinger, and W. Kinzelbach. *Transport of a decay chain in homogeneous porous media: Analytical solutions.* Journal of Contaminant Hydrology, 49(3–4):217–239, 2001.

[22] R.M. Beam and R. F. Warming. *Alternating direction implicit methods for parabolic equations with a mixed derivative.* SIAM J. Sci. Stat. Comput., 1:131–159, 1980.

[23] J. Bear. *Dynamics of Fluids in Porous Media.* American Elsevier, New York, 1972.

[24] J. Bear and Y. Bachmat. *Introduction to Modeling of Transport Phenomena in Porous Media.* Kluwer Academic Publishers, Dordrecht, Boston, London, 1991.

[25] J. Bey. *Finite-Volumen- und Mehrgitter-Verfahren für elliptische Randwertprobleme.* Advances in Numerical Mathematics, B.G. Teubner Stuttgart, Leipzig, 1998 .

[26] M. Bjorhus. *Operator splitting for abstract Cauchy problems.* IMA Journal of Numerical Analysis, 18:419–443, 1998.

[27] D. Braess. *Finite Elemente.* Springer-Verlag, Berlin, Heidelberg, New York, 1992.

[28] S.C. Brenner and C. Carstensen. *Finite Element Methods.* Encyclopedia of Computational Mechanics, John Wiley and Sons, 2004.

[29] S.C. Brenner and L.R. Scott. *The Mathematicla Theory of Finite Element Methods.* Texts in Applied Mathematics, Vol. 15, Springer Verlag, Berlin, Heidelberg, 2002.

[30] J. Budd and A. Iserles. *Geomertic integration: Numerical solution of differential equations on manifolds.* R. Soc. London Philos. Trans. A, 357:943–1133, 1999.

[31] J. Budd and M.D. Piggott. *Geometric integration and its applications.* Lecture notes available, [http://www.bath.ac.uk/ mascjb/home.html], 2002.

[32] D. Buhmann. *Das Programmpaket EMOS. Ein Instrumentarium zur Analyse der Langzeitsicherheit von Endlagern.* Gesellschaft für Anlagen- und Reaktorsicherheit (mbH), GRS-159, Braunschweig, 1999.

[33] J.C. Butcher. *Implicit Runge-Kutta processes.* Math. Comp., 18:50–64, 1964.

[34] J.C. Butcher. *Numerical Methods for Ordinary Differential Equations.* John Wiley & Sons Ltd, Chichester, 2003.

[35] X.C. Cai. *Additive Schwarz algorithms for parabolic convection-diffusion equations.* Numer. Math., 60(1):41–61, 1991.

[36] X.C. Cai. *Multiplicative Schwarz methods for parabolic problems.* SIAM J. Sci Comput., 15(3):587–603, 1994.

[37] S. Candel. *A review of numerical methods in acoustic wave propagation.* in Recent Advances in Aeroacoustics, Springer, New York, 339–410, 1986.

[38] H.S. Carlslaw and J.C. Jäger. *Conduction of Heat in Solids.* Clarendon Press, Oxford, 1959.

[39] C. Carstensen and S.A. Sauter. *A posteriori error analysis for elliptic PDEs on domains with complicated structures.* Numer. Math., 96:691–721, 2004.

[40] M.A. Celia, J.S. Kindred, and I. Herrera. *Contaminant transport and biodegradation: A numerical model for reactive transport in porous media.* Water Resources Research, 25:1141–1148, 1989.

[41] Q.-S. Chen, H. Zhang, V. Prasad, C.M. Balkas, and N.K. Yushin. *Modeling of heat transfer and kinetics of physical vapor transport growth of silicon carbide crystals.* Transactions of the ASME, Journal of Heat Transfer, 123(6):1098–1109, 2001.

[42] S.A. Chin. *A fundamental theorem on the structure of symplectic integrators.* Physics Letters A, 354:373–376, 2006.

[43] W. Cheney. *Analysis for Applied Mathematics.* Graduate Texts in Mathematics, 208, Springer, New York, Berlin, Heidelberg, 2001.

[44] P.G. Ciarlet. *The Finite Element Method for Elliptic Problems.* North Holland, Amsterdam, New York, Oxford, 1978.

[45] P.G. Ciarlet and J.L. Lions (eds.). *Finite Difference Methods (Part 1).* Handbook of numerical analysis, Vol. I, North-Holland/Elsevier, Amsterdam, The Netherlands, 1990.

[46] P.G. Ciarlet and J.L. Lions (eds.). *Finite Element Methods (Part 1).* Handbook of Numerical Analysis, Vol. II, North-Holland/Elsevier, Amsterdam, The Netherlands, 1991.

[47] K.H. Coats and B.D. Smith. *Dead-end pore volume and dispersion in porous media.* Society of Petroleum Engineers Journal, 4(3):73–84, 1964.

[48] B. Cockburn and C.-W. Shu. *The local discontinuous Galerkin method for time-dependent convection-diffusion systems.* SIAM Journal of Numerical Analysis, 35:2240-2463, 1998.

[49] B. Cockburn. *Discontinuous Galerkin Methods for Convection-Dominated Problems.* Higher-Order Methods for Computational Physics, Lecture Notes in Computational Science and Engineering, 9:69–225, 2003.

[50] Gary C. Cohen *Higher-Order Numerical Methods for Transient Wave Equations.* Springer Verlag, Berlin, 2002.

[51] R. Courant, K.O. Friedrichs, and H. Lewy. Collatz. *Über die partiellen Differenzengleichungen der mathematischen Physik.* Math. Ann., 100:32–74, 1928.

[52] M.G. Crandall and A. Majda. *The method of fractional steps for conservation laws.* Math. Comp., 34:285–314, 1980.

[53] M.G.Crandall and A. Majda. *Monotone differences approximations for scalar conservation laws.* Math. Comp., 34:1–21, 1980.

[54] D. Daoud and J. Geiser. *Overlapping Schwarz waveform relaxation for the solution of coupled and decoupled system of convection-diffusion-reaction equation.* Applied Mathematics and Computation, 190(1): 946–964, 2007.

[55] S.M. Day et al. *Test of 3D elastodynamic codes: Final report for lifelines project 1A01.* Technical report, Pacific Earthquake Engineering Center, 2001.

[56] St.M. Day et al. *Test of 3D elastodynamic codes: Final report for lifelines project 1A02.* Technical report, Pacific Earthquake Engineering Center, 2003.

[57] C. N. Dawson, Q. Du, and D.F. Dupont. *A finite difference domain decomposition algorithm for numerical solution of the heat equation.* Mathematics of Computation, 57:63–71, 1991.

[58] J. Douglas, Jr.. *On the numerical integration of $\partial^2 u/\partial x^2 + \partial^2 u/\partial y^2 = \partial u/\partial t$ by implicit methods.* J. SIAM, 3:42–65, 1955.

[59] J. Douglas, Jr. and S. Kim. *Improved accuracy for locally one-dimensional methods for parabolic equations.* Mathematical Models and Methods in Applied Sciences, 11:1563–1579, 2001.

[60] F. Dupret, P. Nicodéme, Y. Ryckmans, P. Wouters, and M.J. Crochet. *Global modelling of heat transfer in crystal growth furnaces.* Intern. J. Heat Mass Transfer, 33(9):1849–1871, 1990.

[61] G. Eason and J. Fulton, and I. N. Sneddon. *The Generation of Waves in an Infinite Elastic Solid by Variable Body Forces.* Phil. Trans. R. Soc. Lond., 1956.

[62] G.R.. Eykolt. *Analytical solution for networks of irreversible first-order reactions.* Wat.Res., 33(3):814–826, 1999.

[63] K.-J. Engel and R. Nagel, *One-Parameter Semigroups for Linear Evolution Equations.* Springer, New York, 2000.

[64] K. Erikson and C. Johnson. *Error estimates and automatic time step control for nonlinear parabolic problems, I.* SIAM Journal on Numerical Analysis archive, 24(1):12–23, 1987.

[65] L.C. Evans. *Partial Differential Equations.* Graduate Studies in Mathematics, Vol. 19, AMS, 1998.

[66] R.E. Ewing. *Up-scaling of biological processes and multiphase flow in porous media.* IIMA Volumes in Mathematics and Its Applications, Springer-Verlag, 295:195–215, 2002.

[67] G. Fairweather and A.R. Mitchell. *A high accuracy alternating direction method for the wave equations.* J. Industr. Math. Appl., 1:309–316, 1965.

[68] I. Farago. *Splitting methods for abstract Cauchy problems.* Lect. Notes Comp. Sci. 3401, Springer Verlag, Berlin, 35-45, 2005.

[69] I. Farago. *Modified iterated operator splitting method.* Applied Mathematical Modeling, 32(8):1542-1551, 2008

[70] I. Farago and J. Geiser. *Iterative Operator-Splitting methods for Linear Problems.* International Journal of Computational Science and Engineering, 3(4):255–263, 2007.

[71] I. Farago and A. Havasi. *Consistency analysis of operator splitting methods for C_0-semigroups.* Semigroup Forum, Springer-Verlag, New York, 74(1):125–139. 2007

[72] E. Fein and A. Schneider. *d^3f- Ein Programmpaket zur Modellierung von Dichteströmungen.* Abschlussbericht, Braunschweig, 1999.

[73] E. Fein, T. Kühle, and U. Noseck. *Entwicklung eines Programms zur dreidimensionalen Modellierung des Schadstofftransportes.* Fachliches Feinkonzept, Braunschweig, 2001.

[74] E. Fein. *Beispieldaten für radioaktiven Zerfall.* Private communications, Braunschweig, 2000.

[75] E. Fein. *Physikalisches Modell und mathematische Beschreibung.* Private communications, Braunschweig, 2001.

[76] P. Frolkovič and J. Geiser. *Numerical Simulation of Radionuclides Transport in Double Porosity Media with Sorption.* Proceedings of Algorithmy 2000, Conference of Scientific Computing, 28–36, 2000.

[77] M.J. Gander and H. Zhao. *Overlapping Schwarz waveform relaxation for parabolic problems in higher dimension.* Proceedings of Algoritmy 1997, Conference of Scientific Computing, 42–51, 1997.

[78] M.J. Gander and S. Vanderwalle. *Analysis of the parareal time-parallel time-integration method.* SIAM Journal on Scientific Computing, 29(2):556–578, 2007.

[79] M.J. Gander and E. Hairer. *Nonlinear Convergence Analysis for the Parareal Algorithm.* Proceedings of the 17th International Conference on Domain Decomposition Methods, 2006.

[80] J. Geiser. *Numerical Simulation of a Model for Transport and Reaction of Radionuclides.* Proceedings of the Large Scale Scientific Computations of Engineering and Environmental Problems, Sozopol, Bulgaria, 2001.

[81] J. Geiser. *Gekoppelte Diskretisierungsverfahren für Systeme von Konvektions-Dispersions-Diffusions-Reaktionsgleichungen.* PhD thesis, University of Heidelberg, 2004.

[82] J. Geiser. R^3T : *Radioactive-Retardation-Reaction-Transport-Program for the Simulation of radioactive waste disposals.* Technical report, Institute for scientific computation, Texas A&M University, College Station, April 2004.

[83] J. Geiser. *Discretisation methods with embedded analytical solutions for convection dominated transport in porous media.* Lect. Notes in Mathematics, Springer-Verlag, Berlin, 3401:288–295, 2005.

[84] J. Geiser. *Discretisation methods with embedded analytical solutions for convection-diffusion dispersion reaction-equations and applications,* J. Eng. Math., 57:79–98, 2007.

[85] J. Geiser. *Mixed discretization methods for the discontinuous Galerkin method with analytical test-functions.* Preprint no. 2006-8 of the Humboldt Universität zu Berlin, Department of Mathematics, Germany, May 2006.

[86] J. Geiser and J. Gedicke. *Iterative operator-splitting methods with higher order time-integration methods and applications for parabolic partial differential equations.* Humboldt-Preprint, no. 2006-10, 2006.

[87] J. Geiser and J. Gedicke. *Nonlinear iterative operator-splitting methods and applications for nonlinear parabolic partial differential equations.* Preprint no. 2006-17 of Humboldt University of Berlin, Department of Mathematics, Germany, 2006.

[88] J. Geiser and Chr. Kravvaritis. *A domain decomposition method based on iterative operator-splitting method.* Preprint of Humboldt University of Berlin, Department of Mathematics, Germany, no. 2006-22, 2006.

[89] J. Geiser and Chr. Kravvaritis. *Weighted Iterative Operator-Splitting Methods for stiff problems and applications.* Preprint No. 2006-11 of Humboldt University of Berlin, Department of Mathematics, Germany, 2006.

[90] J. Geiser. *Weighted Iterative Operator-Splitting Methods: Stability-Theory.* Lecture Notes in Computer Science, 4310:40–47, 2007.

[91] J. Geiser and Chr. Kravvaritis. *Weighted Iterative Operator-Splitting Methods and Applications.* Lecture Notes in Computer Science, 4310:48–55, 2007.

[92] J. Geiser. *Modified Jacobian Newton Iterative Method with Embedded Domain Decomposition Method.* Mathematical Problems in Engineering, Hindawi Publishing Corp., New York, accepted, December 2008.

[93] J. Geiser and L. Noack. *Operator-Splitting Methods Respecting Eigenvalue Problems for Nonlinear Equations and Application in Burgers-Equations.* Preprint 2008-13, Humboldt University of Berlin, Department of Mathematics, Germany, 2008.

[94] J. Geiser, R.E. Ewing, and J. Liu. *Operator Splitting Methods for Transport Equations with Nonlinear Reactions.* Proceedings of the Third MIT Conference on Computational Fluid and Solid Mechanic, Cambridge, MA, June 14-17, 2005.

[95] J. Geiser, O. Klein, and P. Philip. *WIAS-HiTNIHS: Software-tool for simulation in sublimation growth for SiC single crystal: Application and Methods.* Proceeding, IANANO, Houston, Texas, November 2004.

[96] J. Geiser, O. Klein, and P. Philip. *Numerical simulation of heat transfer in materials with anisotropic thermal conductivity: A finite volume scheme to handle complex geometries.* Preprint no.1033, Weierstrass-Institut fr Angewandte Analysis und Stochastik, Berlin, 2005.

[97] J. Geiser, O. Klein, and P. Philip. *Transient numerical study of temperature gradients during sublimation growth of SiC: Dependence on apparatus design.* Journal of Crystal Growth, 297:20–32, 2006.

[98] J. Geiser, O. Klein, and P. Philip. *Influence of anisotropic thermal conductivity in the apparatus insulation for sublimation growth of SiC: Numerical investigation of heat transfer.* Crystal Growth & Design, 6:2021–2028, 2006.

[99] J. Geiser, O. Klein, and P. Philip. *Numerical simulation of temperature fields during the sublimation growth of SiC single crystals, using WIAS-HiTNIHS.* Journal of Crystal Growth, 303(1):352–356, 2007.

[100] J. Geiser and L. Noack. *Iterative operator-splitting methods for wave equations with stability results and numerical examples.* Preprint 2007-10, Humboldt University of Berlin, Department of Mathematics, Germany, 2007.

[101] J. Geiser and L. Noack. *Iterative operator-splitting methods for nonlinear differential equations and applications of deposition processes.* Preprint 2008-04, Humboldt-Universitt zu Berlin, 2008.

[102] J. Geiser and L. Noack. *Operator-splitting methods respecting eigenvalue problems for nonlinear equations and applications for Burgers equations.* Preprint 2008-13, Humboldt University of Berlin, 2008.

[103] J. Geiser and S. Nilsson. *A Fourth Order Split Scheme for Elastic Wave Propagation.* Preprint 2007-08, Humboldt University of Berlin, 2007.

[104] J. Geiser and V. Schlosshauer. *Operator-splitting methods for wave equations.* Preprint 2007-06, Humboldt University of Berlin, Department of Mathematics, Germany, 2007.

[105] J. Geiser and S. Sun. *Multiscale Discontinuous Galerkin Methods for Modeling Flow and Transport in Porous Media.* Lecture Notes in Computer Science, 4487:890–897, 2007.

[106] M.Th. Genuchten. *Convective-Dispersive Transport of Solutes involved in sequential first-order decay reactions.* Computer and Geosciences, 11(2):129–147, 1985.

[107] E. Giladi and H. Keller. *Space time domain decomposition for parabolic problems.* Technical Report 97-4, Center for research on parallel computation CRPC, Caltech, 1997.

[108] R. Glowinski and P. Le Tallec. *Augmented Lagrangian and operator-splitting methods in nonlinear mechanics.* SIAM, studies in applied mathematics, 9, Philadelphia, 1989.

[109] R. Glowinski. *Numerical methods for fluids.* Handbook of Numerical Analysis, Gen. eds. P.G. Ciarlet, J. Lions, Vol. IX, North-Holland Elsevier, Amsterdam, The Netherlands, 2003.

[110] S.K. Godunov. *Difference methods for the numerical calculations of discontinuous solutions of the equations of fluid dynamics.* Mat. Sb., 47:271–306, 1959.

[111] GRAPE. *GRAphics Programming Environment for mathematical problems, Version 5.4.* Institut für Angewandte Mathematik, Universität Bonn und Institut für Angewandte Mathematik, Universität Freiburg, 2001.

[112] W. Hackbusch. *Multi-Grid Methods and Applications.* Springer-Verlag, Berlin, Heidelberg, 1985.

[113] W. Hackbusch. *On first and second order box schemes.* Computing, 41:277–296, 1989.

[114] W. Hackbusch. *Elliptic Differential Equations. Theory and Numerical Treatment* Springer Series in Computational Mathematics, Vol. 18, Springer-Verlag, Berlin, 1992.

[115] W. Hackbusch. *Iterative Solution of Large Sparse Systems of Equations.* Applied Mathematical Sciences, Vol. 95, Springer-Verlag, Berlin, New York, 1994.

[116] E. Hairer, S.P. Norsett, and G. Wanner. *Solving Ordinary Differential Equatons I.* SCM, Springer-Verlag, Berlin, Heidelberg, New York, no. 8, 1992.

[117] E. Hairer and G. Wanner. *Solving Ordinary Differential Equatons II.* SCM, Springer-Verlag, Berlin, Heidelberg, New York, no. 14, 1996.

[118] E. Hairer, C. Lubich, and G. Wanner. *Geometric Numerical Integration: Structure-Preserving Algorithms for Ordinary Differential Equations.* SCM, Springer-Verlag, Berlin, Heidelberg, New York, no. 31, 2002.

[119] A. Harten. *High resolution schemes for hyperbolic conservation laws.* J. Comput. Phys., 49:357–393, 1993.

[120] A. Havasi, J. Bartholy, and I. Farago. *Splitting method and its application in air pollution modeling.* Quarterly Journal of the Hungarian Meteorological Service, 105(1):39–58, 2001.

[121] B. Heinrich. *Finite Difference Methods on Irregular Networks.* International Series of Numerical Mathematics, Birkhäuser, Basel, Boston, Stuttgart, Vol. 82, 1987.

[122] H.McD. Hobgood, M.F. Brady, M.R. Calus, J.R. Jenny, R.T. Leonard, D.P. Malta, St.G. Müller, A.R. Powell, V.F. Tsvetkov, R.C. Glass, and C.H. Carter, Jr.. *Silicon carbide crystal and substrate technology: A survey of recent advances.* Mater. Sci. Forum, 457–460, (2004), 3–8, Proceedings of the 10th International Conference on Silicon Carbide and Related Materials, October 5–10, 2003, Lyon, France.

[123] T.J.R. Hughes. *Multiscale phenomena: Green's functions, the Dirichlet-to-Neumann formulation, subgrid-scale models, bubbles and the origins of stabilized methods* Computer Methods in Applied Mechanics and Engineering, l27:387–401, 1995.

[124] W.H. Hundsdorfer and J. Verwer. *Numerical solution of time-dependent advection-diffusion-reaction equations*, Springer-Verlag, Berlin, 2003.

[125] S. Hu and N.S. Papageorgiou. *Handbook of Multivalud Analysis I,II.* Kluwer, Dordrecht, Part I: 1997, Part II: 2000.

[126] W. Hundsdorfer and L. Portero. A note on iterated splitting schemes. CWI Report MAS-E0404, Amsterdam, Netherlands, 2005.

[127] K. Johannsen. *Robuste Mehrgitterverfahren für die Konvektions-Diffusions Gleichung mit wirbelbehafteter Konvektion.* Doktor-Arbeit, Universität Heidelberg, 1999.

[128] K. Johannsen. *An aligned 3D-finite-volume method for convection-diffusion problems.* Modeling and Computation in Environmental Sciences, R. Helmig, W. Jäger, W. Kinzelbach, P. Knabner, and G. Wittum (eds.), Vieweg, Braunschweig, 59:227–243, 1997.

[129] W.A. Jury, K. Roth. *Transfer Functions and Solute Movement through Soil.* Bikhäuser Verlag Basel, Boston, Berlin, 1990.

[130] J. Kanney, C. Miller, and C.T. Kelley. *Convergence of iterative split-operator approaches for approximating nonlinear reactive transport problems.* Advances in Water Resources, 26:247–261, 2003.

[131] C.T. Kelley. *Solving Nonlinear Equations with Newton's Method.* Computational Mathematics, SIAM, XIV, 2003.

[132] C.A. Kennedy and M.H. Carpenter. *Additive Runge-Kutta schemes for convection-diffusion-reaction equations.* Applied Numerical Mathematics, 44(1–2):139–181, 2003.

[133] S. Kim and H. Lim. *High-order schemes for acoustic waveform simulation.* Applied Numerical Mathematics, 57(4):402–414, 2007.

[134] J. Herzer and W. Kinzelbach. *Coupling of transport and chemical processes in numerical transport models.* Geoderma, 44:115–127, 1989.

[135] W. Kinzelbach. *Numerische Methoden zur Modellierung des Transports von Schadstoffen im Grundwasser.* Schriftenreihe Wasser-Abwasser, Oldenburg, 1992.

[136] O. Klein and P. Philip. *Transient numerical investigation of induction heating during sublimation growth of silicon carbide single crystals.* J. Crystal Growth, 247(1-2):219–235, 2003.

[137] O. Klein and P. Philip. *Transient temperature phenomena during sublimation growth of silicon carbide single crystals.* J. Crystal Growth, 249(3–4):514–522, 2003.

[138] O. Klein, P. Philip, and J. Sprekels. *Modeling and simulation of sublimation growth of SiC bulk single crystals.* Interfaces and Free Boundaries, 6:295–314, 2004.

[139] O. Klein, P. Philip, J. Sprekels, and K. Wilmański. *Radiation- and convection-driven transient heat transfer during sublimation growth of silicon carbide single crystals.* J. Crystal Growth, 222(4):832–851, 2001.

[140] A.O. Konstantinov. *Sublimation growth of SiC*, G.L. Harris (ed.), Properties of Silicon Carbide, EMIS Datareview Series, no. 13, London, UK, 170–203, 1995.

[141] R. Kozlov and B. Owren. *Order reduction in operator splitting methods.* Preprint N6-1999, Department of Mathematical Sciences, Norwegian University of Science and Technology, Trondheim, Norway, 1999.

[142] R. Kozlov, A. Kvarno, and B. Owren. *The bahaviour of the local error in splitting methods applied to stiff problems.* Journal of Computational Physics, 195:576–593, 2004.

[143] D. Kröner. *Numerical Schemes for Conservation Laws.* Wiley and Teubner, Chichester, New York, 1997.

[144] L.D. Landau and E.M. Lifschitz. *On the theory of the dispersion of magnetic permeability in ferromagnetic bodies.* Collected papers of L.D. Landau (D.ter. Haar eds.), pp. 101–114, Pergamon, 1965.

[145] D. Lanser and J.G. Verwer. *Analysis of operator-splitting for advection-diffusion-reaction problems from air pollution modelling.* Journal of Computational Applied Mathematics, 111(1–2):201–216, 1999.

[146] S. Larsson and V. Thomee. *Partial Differential Equations with Numerical Methods.* Text in Applied Mathematics 45, Springer-Verlag, Berlin, Heidelberg, 2003.

[147] P. Lax. *Hyperbolic Systems of Conservation Laws and the Mathematical Theory of Shock Waves.* SIAM, 1973.

[148] P. Lax. *Functional Analysis.* Wiley-Interscience, New York, 2002.

[149] M. Lees. *Alternating direction methods for hyperbolic differential equations.* J. Soc. Industr. Appl. Math., 10(4):610–616, 1962.

[150] R.J. LeVeque. *Finite Volume Methods for Hyperbolic Problems.* Cambridge Texts in Applied Mathematics, Cambridge University Press, 2002.

[151] C. Lubich. *A variational splitting integrator for quantum molecular dynamics.* Applied Numerical Mathematics archive, 48(3–4):355–368, 2004

[152] R. Madar, J. Camassel, and E. Blanquet. *Silicon Carbide and Related Materials ICSCRM, Lyon, France, October 5–10, 2003.* Materials Science Forum, Vol. 457–460, Part II, Trans Tech Publications Ltd., 2004.

[153] R. März. *Numerical methods for differential-algebraic equations.* Acta Numerica, 141–198, 1992.

[154] G.I. Marchuk. *Some applications of splitting-up methods to the solution of mathematical physics problems.* Aplikace Matematiky, 13:103–132, 1968.

[155] R.I. McLachlan, G. Reinoult, and W. Quispel. *Splitting methods.* Acta Numerica, 341–434, 2002.

[156] J.M. Melenk and I. Babuska. *The partition of unity finite element method: Theory and Application.* Comput. Methods Appl. Mech. Engrg., 139:289–314, 1996.

[157] G.A. Meurant. *Numerical experiments with a domain decomposition method for parabolic problems on parallel computers.* In R. Glowinski, Y.A. Kuznetsov, G.A. Meurant, J. Périaux, and O. Widlund, editors, *Fourth International Symposium on Domain Decomposition Methods for Partial Differential Equations*, SIAM, Philadelphia, PA, 1991.

[158] J. Meziere, M. Pons, L. Di Cioccio, E. Blanquet, P. Ferret, J.M. Dedulle, F. Baillet, E. Pernot, M. Anikin, R. Madar, and T. Billon. *Contribution of numerical simulation to silicon carbide bulk growth and epitaxy.* J. Phys.-Condes. Matter, 16:1579–1595, 2004.

[159] St.G. Müller, R.C. Glass, H.M. Hobgood, V.F. Tsvetkov, M. Brady, D. Henshall, D. Malta, R. Singh, J. Palmour and C.H. Carter Jr.. *Progress in the industrial production of SiC substrate for semiconductor devices.* Mater. Sci. Eng. ,B 80(1–3):327–331, 2002.

[160] F. Neri. *Lie algebras and canonical integration.* Department of Physics, University of Maryland, Preprint, 1988.

[161] N. Neuss. *A new sparse matrix storage methods for adaptive solving of large systems of reaction-diffusion-transport equations.* In Keil et.al., ed., Scientific Computing in Chemical Engineering II, Springer-Verlag, Berlin, Heidelberg, New York, 175–182, 1999.

[162] Website: OFELI: http://ofeli.sourceforge.net/

[163] M. Ohlberg. *A posteriori error estimates for vertex centered finite volume apprximations of convection-diffusion-reaction equations.* Preprints 12/2000, Department of Mathematics, University of Freiburg, Germany, May 2000.

[164] C.V. Pao. *Non Linear Parabolic and Elliptic Equation.* Plenum Press, New York, 1992.

[165] M. Parashar, V. Matossian, W. Bangerth, H. Klie, B. Rutt, T.M. Kurc, U.V. Catalyuerek, J.H. Saltz, and M.F. Wheeler. *Towards dynamic data-driven optimization of oil well placement.* International Conference on Computational Science, 2:656–663, 2005.

[166] N.H. Pavel. *Nonlinear evolution operators and semigroups.* Lect. Notes in Math., 1260, Springer, Berlin, 1987.

[167] D.W. Peaceman and H.H. Racheford. Jr.. *The numerical solution of parabolic and elliptic differential equations.* J. SIAM, 3:28–42, 1955.

[168] J. Peraire and P.-O. Persson. *A compact discontinuous Galerkin (CDG) method for elliptic problems.* SIAM Journal on Scientific Computing, 30(4), 1806–1824, 2008.

[169] P. Philip. *Transient numerical simulation of sublimation growth of SiC bulk single crystals: Modeling, finite volume method, results.* Ph.D. thesis, Department of Mathematics, Humboldt University of Berlin, Germany, 2003, Report no. 22, Weierstra-Institut fr Angewandte Analysis und Stochastik, Berlin.

[170] L. Portero. *Fractional step Runge-Kutta methods for multidimensional evolutionary problems with time-dependent coefficients and boundary conditions.* PhD-thesis, Universidad Publica de Navarra, 2007.

[171] L. Portero, B. Bujanda, and J.C. Jorge. *A combined Embedded Fractional Step Domain Decomposition operator splittings for the solution of parabolic equations.* Springer Lecture Notes in Comput. Sci., 3019:1034–1041, 2004.

[172] L. Portero and J.C. Jorge. *Embedded pairs of fractional step Runge-Kutta methods and improved domain decomposition techniques for*

parabolic problems. Springer Lecture Notes in Computational Science and Engineering, 55:731–738, 2007

[173] A. Prechtel. *Modelling and Efficient Numerical Solution of Hydrogeochemical Multicomponent Transport Problems by Process-Preserving Decoupling Techniques.* PhD-Thesis, University of Erlangen, Germany, 2005.

[174] A. Prohl. *Computational Micromagnetism.* Advances in Numerical Mathematics, Teubner, Stuttgart, 2001.

[175] G.F. Roach. *Green's Functions.* Cambridge University Press, 1970.

[176] M. Rumpf and A. Wierse. *GRAPE, Eine interaktive Umgebung für Visualisierung und Numerik.* Informatik, Forschung und Entwicklung, 1990.

[177] T.F. Russell and M.A. Celia. *An overview of research on Eulerian-Lagrangian localized adjoint methods (ELLAM).* Advances in Water Resources, 25:1215–1231, 2002.

[178] A. Quarteroni and A. Valli. *Numerical Approximation of Partial Differential Equations.* Springer Series in Computational Mathematics, Springer-Verlag, Berlin, Heidelberg, New-York, 1997.

[179] A. Quarteroni and A. Valli. *Domain decomposition methods for partial differential equations* Series: Numerical Mathematics and Scientific Computation, Clarendon Press, Oxford, 1999.

[180] H.A. Schwarz. *Über einige Abbildungsaufgaben.* Journal für Reine und Angewandte Mathematik, 70:105–120, 1869.

[181] H.R. Schwarz. *Numerische Mathematik.* B.G. Teubner Verlag Stuttgart, 2. Auflage, 1988.

[182] H.R. Schwarz and J. Waldvogel. *Numerical Analysis. A Comprehensive Introduction.* John Wiley and Sons, Chichester, 1989.

[183] Th. Sonar. *On the design of an upwind scheme for compressible flow on general triangulation.* Numerical Analysis, 4:135–148, 1993.

[184] B. Sportisse. *An analysis of operator-splitting techniques in the stiff case.* Journal of Computational Physics, 161:140–168, 2000.

[185] G. Strang. *On the construction and comparision of difference schemes.* SIAM J. Numer. Anal., 5:506–517, 1968.

[186] G. Strang and G.J. Fix. *An Analysis of the Finite Element Method.* Prentice-Hall Series in Automatic Computation, Prentice-Hall, Inc., Englewood Cliffs, N.J., 1973.

[187] V. Thomee. *Galerkin Finite Element Methods for Parabolic Problems.* Lecture Notes in Math., vol. 1054, Springer-Verlag, Berlin, New York, Heidelberg, 1984.

[188] A.-K. Tornberg and B. Engquist. *Numerical approximations of singular source terms in differential equations.* J. Comput. Phys., 200(2):462–488, 2004

[189] A. Toselli and O. Widlund. *Domain Decomposition Methods – Algorithms and Theory* Series: Springer Series in Computational Mathematics, Vol. 34, 2005.

[190] L.N. Trefethen and M. Embree. *Spectra and Pseudospectra: The Behaviour of Nonnormal Matrices and Operators.* Princeton University Press, Princeton, New Jersey, 2005.

[191] V.S. Varadarajan. *Lie-Groups, Lie Algebra and Their Representations.* Prentice Hall, Englewood Cliffs, New Jersey, 1974.

[192] R. Verfürth. *A Review of a posteriori Error Estimates and Adaptive Mesh-Refinement Techniques.* Wiley and Teubner, Chichester, New York, 1996.

[193] J.G. Verwer and B. Sportisse. *A note on operator splitting in a stiff linear case.* CWI, Amsterdam, Netherlands, MAS-R9830, 1998.

[194] S. Vandewalle. *Parallel Multigrid Waveform Relaxation for Parabolic Problems.* Teubner Skripten zur Numerik, B.G. Teubner Stuttgart, 1993.

[195] J. Waldén. *On the Approximation of Singular Source Terms in Differential Equations.* Numer. Meth. Part. D E, 15:503–520, 1999.

[196] G. Wittum. *On the robustness of ILU smoothing.* SIAM J. Sci. Statist. Comput., 10:699–717, 1989.

[197] B. Wohlmuth. *Discretisation methods and iterative solvers based on domain decomposition.* Lecture Notes in Computational Science and Engineering, 17, Springer-Verlag, Berlin, Heidelberg, New York, 2003.

[198] K. Yoshida. *Functional Analysis.* Classics in Mathematics, Springer-Verlag, Berlin, Heidelberg, New York, 1980.

[199] H. Yoshida. *Construction of higher order symplectic integrators.* Physics Letters A, 150(5–7):262-268, 1990.

[200] E. Zeidler. *Nonlinear Functional Analysis and Its Applications. II/A Linear Monotone Operators.* Springer-Verlag, Berlin, Heidelberg, New York, 1990.

[201] E. Zeidler. *Nonlinear Functional Analysis and Its Applications. II/B Nonlinear Monotone Operators.* Springer-Verlag, Berlin, Heidelberg, New York, 1990.

[202] Z. Zlatev. *Computer Treatment of Large Air Pollution Models.* Kluwer Academic Publishers, Dordrecht, 1995.

Index